THE ELECTROGENIC SODIUM PUMP

BY

G. A. KERKUT

*Professor of Physiology and Biochemistry,
University of Southampton*

AND

BARBARA YORK

*Department of Physiology and Biochemistry,
University of Southampton*

BRISTOL: SCIENTECHNICA (Publishers) LTD.
1971

COPYRIGHT NOTICE

© SCIENTECHNICA (PUBLISHERS) LTD., 1971

All rights reserved. No part of this publication may be reproduced, stored in a retrieval system, or transmitted in any form or by any means, electronic, photocopying, recording, or otherwise, without the prior permission of the copyright owner.

Distribution by Sole Agents:
United States of America: The Williams & Wilkins Company, Baltimore
Canada: The Macmillan Company of Canada Ltd., Toronto

ISBN 85608 000 4

PRINTED IN GREAT BRITAIN BY JOHN WRIGHT AND SONS LTD., AT THE STONEBRIDGE PRESS, BRISTOL BS4 5NU

PREFACE

THIS book considers the idea that the sodium pump in some muscle and nerve-cells can *directly* contribute to that cell's membrane potential. If the pump is stopped by means of the addition of ouabain, there is a rapid fall in the membrane potential of that cell. The extent to which the sodium pump is electrogenic, i.e., *directly* contributes to the membrane potential, has been under discussion for the past fifteen years. This book presents in Part I an historical account of the subject and indicates some of the reasons why the electrogenic sodium pump's contribution to the membrane potential was discounted. In some ways it is similar to the reluctance in accepting the idea that the ionic imbalance across the cell membrane was due to an energy-consuming system: the sodium pump. We have tried to present some of this story in a general manner in Part I so that the wider implications can be seen.

The remaining Parts II–VII present the evidence in more detail and with fuller documentation for the development of the ideas and experiments. We have concentrated on the evidence from nerve and muscle and have left out the evidence for intestinal cells and kidney cells.

There is also indication that the electrogenic sodium pump is not just a steady system but one that can be varied under physiological conditions. It could play a role at some synaptic sites, in adaptation of some sense organs, and in the relationship between the glial cells and the neurons.

We should like to thank the authors who have given us permission to quote from their works and to use the illustrations. We should also like to thank the Editors and Publishers of *Nature*, *Science*, *Circulation Research*, *Comparative Biochemistry and Physiology*, *Comptes Rendus des Sciences de la Société de Biologie*, the *Journal of Experimental Biology*, the *Journal of General Physiology*, the *Journal of Neurophysiology*, the *Journal of Physiology*,

Verhandlungen der Deutschen zoologischen Gesellschaft, the Clarendon Press, and the Souvenir Press, for permission to reproduce quotations and illustrations.

June, 1971   G. A. K.
                                                                                            B. Y.

CONTENTS

	PAGE
Part I	
THE NATURE OF SCIENTIFIC PROGRESS	1
Part II	
THE ELECTROGENIC SODIUM PUMP IN NERVE AXONS	57
Part III	
THE ELECTROGENIC SODIUM PUMP IN NERVE-CELLS	73
Part IV	
THE ELECTROGENIC SODIUM PUMP IN MUSCLE-FIBRES	107
Part V	
THE SIGNIFICANCE OF THE ELECTROGENIC SODIUM PUMP	131
Part VI	
SOME PROPERTIES OF THE ELECTROGENIC SODIUM PUMP	155
Part VII	
GENERAL CONCLUSION	163
REFERENCES	165
AUTHOR INDEX	175
SUBJECT INDEX	178

ABBREVIATIONS

ACh	Acetylcholine
AP	Action potential
ATP	Adenosine 5′-triphosphate
ATPase	Adenosine 5′-triphosphatase
Ba^{2+}, Ba	Barium ion
$BaCl_2$	Barium chloride
Ca^{2+}, Ca	Calcium ion
CILDA	Cell showing inhibition of long duration
Cl^-, Cl	Chloride ion
cm.	Centimetre
°C.	Degrees centigrade
DNP	2,4-Dinitrophenol
Dopamine	3-Hydroxytyramine
E_K, E_{Na}, E_C	Equilibrium potential for potassium ions, sodium ions, and chloride ions respectively
EPSP	Excitatory post-synaptic potential
g.	Gram
GABA	Gamma aminobutyric acid
H^+	Hydrogen ion
IPSP	Inhibitory post-synaptic potential
I, V	Current, voltage
K^+, K	Potassium ion
KAc	Potassium acetate
Li^+, Li	Lithium ion
LiCl	Lithium chloride
Li_2SO_4	Lithium sulphate
M	Molar
mA.	Milliamp
mM	Millimolar
μM	Micromolar
mm Hg	Millimetres mercury pressure
MP	Membrane potential
mV.	Millivolt
μ	Micron

ABBREVIATIONS

Na^+, Na	Sodium ion
NaAc	Sodium acetate
NaCl	Sodium chloride
Na_2SO_4	Sodium sulphate
P	Inorganic phosphate
P_K P_{Na} P_{Cl}	Permeability of the membrane to potassium, sodium, and chloride
P_{O_2}	Partial pressure of oxygen
PTH	Post-tetanic hyperpolarization
Rb^+, Rb	Rubidium ion
RP	Resting potential
SO_4^{2-}	Sulphate ion
V_m	Membrane potential

MEMBRANE POTENTIAL

NOTE that the resting membrane potential has a negative value. Thus in nerve-cells the membrane potential is approximately −70mV. At times the negative sign has been left out of the text. On all these occasions the potential should be taken as a negative one unless it is specified that it is a positive potential, i.e., +45 mV.

Similarly one may state that the membrane potential has increased in value from −60 mV. to −70 mV.

PART I

INTRODUCTION

THE NATURE OF SCIENTIFIC PROGRESS

Two theories can be given to describe the nature of scientific progress.

STRAIGHT-LINE THEORY

One interpretation of scientific progress is that at a given stage in time somebody carries out a new and critical experiment that indicates quite clearly that a new theory must be correct.

CYCLICAL THEORY

Another interpretation is that at a given time there is a considerable body of experimental evidence supporting view A and another set of data supporting view B. For various reasons view A is accepted as the correct interpretation and the data supporting view B are overruled or ignored. Then a change in the climate of opinion takes place and the data concerning view B are considered to be more valid. View B is then accepted and view A rejected.

The two approaches to scientific advance are not mutually exclusive since view A can be toppled in favour of view B by the advent of new experimental data.

In general, however, it may be suggested that much of the experimental data necessary to provide the new theories of the next five years are already available. These data are being ignored or rejected. The problem is to see the situation afresh and to select the valid data from the invalid material.

To some extent the present book will indicate that the view that cells were permeable to sodium and potassium ions was rejected for ten years, in spite of the evidence already available that cells *were* permeable to sodium and potassium. Perhaps this was because we could not understand how the cell membrane could differentiate between two such similar ions? Similarly, the evidence in favour of the electrogenic sodium pump was rejected because it was more simple to explain membrane potentials without a metabolic component. Over the past five years much of the evidence in favour of the newer view has been available but it has not been considered to be either valid or necessary.

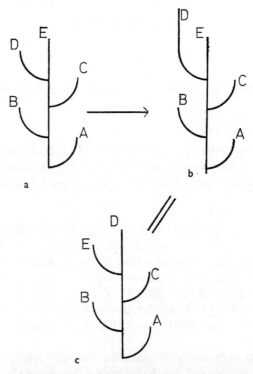

Fig. 1.—Diagram showing the development of theories. **a**, Shows that E is the most accepted theory and that D and C are less accepted. **b**, Shows that D has now become the accepted theory instead of E. **c**, The figure is redrawn so that, although D is the accepted view, it now looks as if it was the natural development and that it was always on the main line instead of once being on a side-line.

A diagrammatic representation of the situation can be illustrated as in *Fig.* 1. The different theories, **A**, **B**, **C**, **D**, etc., can be shown as a tree where the most advanced, view **E**, is the accepted one. a shows this tree with E the most advanced in time. b shows that during the course of time theory D develops and now becomes the accepted view. What most people do now is to redraw the figure as in c so that it seems that a simple straight line of thought has led to D. This is a simplification due to hindsight.

At any one time the situation can be expressed as in *Fig.* 2. We stand at the present time (now) and see that theory C seems to be the correct one. It is possible, however, that any of the other theories might be the future correct ones and it is necessary to look at these rejected theories very carefully to see if A, B, D, or E might become the correct explanation for the future.

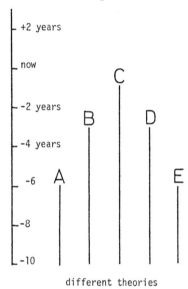

Fig. 2.—The choosing of an adequate theory. At a given time (now) one theory (C) will appear to be the most acceptable. Other theories such as A, B, D, and E have all been considered and there is evidence in their favour but they have been rejected in favour of C. It is possible that the future advance will come from any of A–E and that the evidence is already available.

A similar situation exists with regard to mechanical inventions. The magnetic recorder was devised by Poulsen in 1900, though disk recording was dominant up to the 1950's, after which magnetic recording was redeveloped and became a very important means of recording both in industry and in the home. Similarly, in typewriters the rotating-head machine was developed in 1829, with a commercial model in 1866, but then lost favour to the standard machine with the type in linear array. The rotating-head machine was redeveloped in the 1950's by IBM as the 'golf-ball' machine with its advantages of interchangeable heads and rapid typing speed. In both situations an idea was developed to a limited extent and then left to lie dormant till further technical advances had been made, which then showed the old idea to have certain advantages over the 'current' idea.

One could suggest that research workers, who have to spend much of their time consulting the literature of the past decade, should give special attention to theories or views that are not currently in favour. It is likely that the future advance will be found amongst these rejected ideas and that modern experimental techniques will make it possible to design experiments that will show the *new* theory more clearly and unequivocally than was possible in the past when it was an old, discarded, and forgotten theory. But note: it does not follow that *all* the rejected theories will become correct in time, even in infinite time.

The old story is still true. If you make an original discovery and explain it in a new manner, the reader will say: 'This view is quite impossible and unacceptable. It does not fit in with present or past knowledge.' Then ten years later when everyone has come round to the new way of thinking he will say: 'Of course, there was nothing very new in this idea. It was first suggested by Smith in 1837 and verified by Brown in 1931. We knew it was right all the time and could not see why anyone made a fuss or rejected it.'

Definitions

Three definitions will be given here.

I. Sodium Pump

This is the mechanism by which sodium ions are actively pumped against the sodium concentration gradient.

II. Sodium-potassium linked Pump

1. The pump mechanism is bound to a membrane.
2. Na^+ ions are pumped outwards; K^+ ions are pumped inwards.
3. Three Na ions are pumped out for every 2 K ions pumped in.
4. Energy (ATP) is used for this process.
5. A membrane-bound enzyme (APTase) is involved. It is stimulated by internal Na^+ and external K^+. Mg^{++} is also necessary.

6. The pump can work to a limited extent if there are no K^+ ions in the external medium.
7. The pump can be slowed or inhibited by:—
   a. Removal of external K^+.
   b. Addition of ouabain.
   c. Addition of lithium.
   d. Lowering the temperature.

III. ELECTROGENIC SODIUM PUMP

1. When the pump actively moves Na^+ across the membrane a potential develops across the membrane.
2. This potential falls if the pump is slowed or inhibited. The potential developed is proportional to the activity of the pump.
3. The potential developed depends on the resistance of the membrane.
4. If the membrane has a high resistance then the electrogenic sodium pump can develop a potential of up to -40 mV. across the membrane.
5. If the membrane has a low resistance—due to movement of ions such as chloride across the membrane—then the potential is short-circuited and may be only -1 to -2 mV.
6. The electrogenic sodium pump has a high temperature coefficient.

IS THE ELECTROGENIC SODIUM PUMP IMPORTANT?

The following seven points will indicate where the electrogenic sodium pump might play a role in nerve- and muscle-cells.

MEMBRANE POTENTIAL

1. In many muscle- and nerve-cells the resting membrane potential is explicable in terms of the membrane permeability to potassium ions. The membrane potential is dominated by E_K. There may be a small, say 5 per cent, component of the membrane potential that is due to the activity of the electrogenic sodium pump.
2. In other nerve-cells the contribution of the electrogenic sodium pump to the membrane potential may be quite high. It may contribute 10–60 per cent of the membrane potential. This may be the case in cells that have a high rate of sodium entry or where the surface area/volume ratio is high.

DRUGS

3. The electrogenic sodium pump can be affected by drugs and can rapidly change the membrane potential of a nerve- or muscle-cell and so affect the firing threshold.

TRANSMITTER—INHIBITION

4. Some transmitters such as acetylcholine can affect specific nerve-cells either by acting directly on the electrogenic sodium pump or by allowing sodium ions to leak into the cell and so affect the pump. The resulting hyperpolarization can last for milliseconds, seconds, or minutes and may explain some of the longer-term actions in the central nervous system.

TRANSMITTER—EXCITATION

5. There is evidence that some chemicals can inhibit the electrogenic sodium pump and so lead to a depolarization of the cell.

ANOXIA

6. The electrogenic sodium pump is sensitive to anoxia. Reducing the partial pressure of oxygen slows or stops the pump and leads to a fall in the membrane potential and a change in the cell excitability.

GENERATOR POTENTIAL, ADAPTATION, GLIAL CONTROL, AND CHROMATOPHORES

7. The electrogenic sodium pump may play a role in the potential of some sense organs, allow adaptation of sense organs, allow the control of nerve-cells by glial cells, and contribute to the control of chromatophores.

HISTORICAL INTRODUCTION

In this historical introduction some of the data concerning the development of the idea that cells could have a high internal potassium concentration and a low sodium concentration will be presented. It will indicate the manner in which specific ideas can dominate a subject for many years before they are finally modified or rejected and discarded. It is a case history for *Fig.* 1.

Studies of the chemical and ionic composition of cells by early workers, such as Hamburger (1891), Overton (1895), and Gurber

(1895), showed that the composition of the cytoplasm within the cells differed from the ionic composition of the plasma or extracellular fluid. In general it appeared that the cells had a higher concentration of potassium and a lower concentration of sodium and chloride inside the cells than was present in the plasma. These early studies raised the question: 'How can the cell contain a higher concentration of potassium ions than is present in the plasma, especially if the cell membrane is permeable to potassium ions?'

To appreciate the problem and its answers, one must consider the state of physiology and biochemistry in the early part of the twentieth century. Although there had previously been a uniformity in the approach to sciences and a generally educated man such as Herbert Spencer found it possible to study science as well as philosophy, religion, and literature, by the onset of the twentieth century there was a change and the biologists were having a struggle to establish themselves as 'accepted scientists'. The sciences of chemistry and physics and the techniques of mathematics were dominant in the hierarchical system, but the biological sciences were still under the shadow of vitalism and theology. Many biologists felt that they had, above all other things, to explain biological problems in terms of physics and chemistry and, what is more, in the terms of the physics and chemistry *known at that time*.

The biologist could not suggest that the internal potassium concentration within the cell was high because God had made it so, or because it had evolved in that way, or because there was a special vital force that had put the potassium there. In order to be accepted within the body of contemporary science the biologist had to explain the high concentration of potassium within the cell in terms of the law of mass action or some simple physicochemical system that most chemists or physicists would accept as valid or obviously true.

An indication of the dilemma facing the student of biological systems can be seen from the following example. J. S. Haldane had been interested for many years in the transport of gases through lungs and tissues. In particular he had been concerned with the physiology of respiration but he also studied the secretion of gas into the swim-bladder of fishes. Haldane was very surprised when he found that fish that lived at considerable depths in the ocean had air in their swim-bladders since he realized that, according to Henry's law, the high pressure should have forced the gas into solution. It was clearly against Henry's law that the swim-bladder should contain gas. The biological phenomenon was not explicable in terms of physics and chemistry. Haldane (1935) realized that scientific knowledge was at that time limited and he stated: 'In reality there is not the slightest scientific indication that we shall ever be able in physiology to dispense with the specific conception of life.... It is often argued that if we place biology on a different theoretical footing from physics and chemistry we are thereby abandoning attempts to explain the phenomena of life, and that however unsatisfactory these attempts may have proved we can do nothing better than continue them. The reply to this criticism is that an unjustifiable philosophical assumption is implied in the conception that we must be able to interpret in terms of physical science the universe which we perceive around us.... Existing physical science can give no account of the characteristic features of life and conscious experience or their assumed origin in the course of evolution.'

To some extent J. S. Haldane was a vitalist because he did not feel that physics and chemistry could explain all the phenomena he was studying. The case of the gas in the swim-bladder was fairly simply explained in more recent times by studies on the enzyme carbonic anhydrase which facilitates the reaction $CO_2 + H_2O = H_2CO_3$. This enzyme can displace carbon dioxide from carbonic acid and the equilibrium constant is such that it requires several hundred atmospheres' pressure to push the gas back into solution. This reaction is partially involved in producing the gas in the swim-bladder of fish. We now have an explanation of the manner in which Henry's law is apparently disobeyed. The physics and chemistry of Haldane's time were not sufficient since they did not fully indicate the manner in which this enzyme could powerfully alter the equilibrium point of the reaction, though it is interesting to note that Haldane was aware of some of the work on this enzyme.

In the present book we shall assume that the problems of ionic imbalance across membrane and the problems of the origins of the potentials across membrane will ultimately be fully explained in terms of physics and chemistry. It may require the physics and chemistry of the twenty-fifth century to do so (and in this we are indicating an act of faith), but we think that in time it should be possible for us to build artificial systems that mimic and use the same basic principles of physics, chemistry, and engineering as those used in living systems.

An indication of a modern physicist's view is given by Kapitsa in his articles on 'The Future Problems of Science' (1964): 'I think that we physicists hold an incorrect view of biology. We look on it from the standpoint that we can interpret biological processes on the basis of physical processes. I feel that the real position is precisely the opposite. Studying biology we must enrich physics. Nature is still a better engineer–builder than men, and we still do not understand all it has built and performed. It knows the laws of nature better than man, and our relation to physics and biology must change in connexion with present-day opinions. We physicists must learn the laws of nature. In truth, we have overtaken Nature in a certain sense; for example, in the chain reaction of uranium fission; Nature did not create it, but man, yes, so that we can compete with Nature. This is the same in muscle power; Nature transforms chemical power into muscle power, but we do not understand this process as yet.'

CHEMICAL ANALYSIS OF CYTOPLASM

Two technical problems faced research workers interested in the differences in ionic concentration between the inside and the outside of the cell. The first difficulty was choice of a suitable method of chemical analysis. Early workers tended to use gravimetric and titrometric methods of analysis. These were often very time-consuming and subject to considerable experimental error.

The advent of colorimetric methods, flame photometry, and the use of radioactive tracers put the problems of chemical analysis in new perspective and it became possible to do in days what had previously taken weeks or months to do. The feeling is clearly expressed by Wallace Fenn (1962) in his article 'Born Fifty Years Too Soon': 'A modern flame photometer calls dismally to mind the years wasted in precipitating and titrating those hundreds of sodium and potassium samples. More time might have been spent on technical advances before trying to work with inadequate equipment. The manpower working in the physiology of electrolytes today could finish in one year all the work done by their relatively few predecessors between 1900 and 1940 and the results would be better.'

The Red Blood-corpuscle

A second difficulty is in obtaining pure cytoplasm for analytical purposes since too often the sample will be contaminated with extracellular fluid or serum. In the case of the red blood-corpuscle it is fairly easy to estimate the amount of extracellular fluid spun down with the corpuscles since one can add an indicator such as inulin and estimate this. If one centrifuges the red blood-corpuscles for a standard time of 30 minutes at 1500 g then the packed cells contain 3–5 per cent of trapped plasma.

The values for the Na and K levels in the red blood-corpuscles of different species are shown in Table I. There are clearly two different classes of red blood-corpuscles: those with a higher

Table I.—The Approximate Concentration of Sodium and Potassium in Red Blood-cells*

Subject	Sodium (mEq./l.)	Potassium (mEq./l.)
Man	19	136
Rabbit	22	142
Rat	28	135
Horse	16	140
Sheep	98	46
Ox	104	35
Cat	142	8
Dog	135	10

* Note the two groups of cells, i.e., low and high potassium content.

internal potassium concentration and those with a lower internal potassium concentration. Even in the case of the sheep, ox, cat, and dog, where the internal potassium concentration of the red blood-corpuscle is low and the sodium concentration high, the sodium level is still less than that present in the plasma.

In the sheep it is possible to find some breeds that have red blood-cells with a high potassium content and other breeds with a low potassium content. It is of interest to note that there is a correlation between the type of red blood-cell and the level of ATPase in the membrane. Those animals that have a higher level of ATPase in the membrane have the higher internal K^+.

Sodium and Potassium Levels in Muscle

An estimation can be made of the amount of extracellular space between the muscle-fibres by treating the system with inulin which will not penetrate into the muscle-cells. Measurement of potassium levels in muscle samples taken during operations (muscle biopsy) gave consistent results on rats where 30-mg. samples proved to be sufficient. In man, however, it was necessary to take larger samples (Flear, 1962; Flear and Florence, 1963). In general it is possible to take quite large samples of muscles and estimate the concentration of sodium and potassium within the cells after making the necessary correction for the contamination by the extracellular fluid. Table II shows the approximate

Table II.—The Approximate Concentration* of Ions in the Muscles and Plasma of Selected Animals

Subject	In Muscle		In Plasma	
	K	Na	K	Na
Man	80	12	4·5	135
Dog	140	12	4·0	150
Rat	140	13	4·0	150
Frog	125	15	2·6	110
Crab	188	24	7·6	518
Lobster	152	79	12·5	502

* Concentrations are given in mEq./l.

levels of the ions in the muscles of selected animals, together with the plasma levels.

In general all muscles have a high potassium content and a low sodium content. The levels differ according to the animal species selected and there are some indications that there may be, in some invertebrate animals, specific differences between muscles of the same animal.

Sodium and Potassium Levels in Nerve

The simplest method of obtaining nerve axoplasm occurs in the squid giant axon where the single axons are up to 1 mm. in diameter and it is easy to obtain 2-cm. lengths of axon. The axoplasm can then be squeezed out and obtained relatively free from the contaminating extracellular fluid (Koechlin, 1955). In the case of crab nerve it is necessary to dissect out single axons which also contain small amounts of glial tissue (Keynes and Lewis, 1951). In mammalian nerve the situation is more difficult since the axons run in nerve-trunks which contain a relatively large amount of extracellular fluid. It is necessary to make quite a large correction for this. The problem of the size of this correction for the extracellular space is shown in the estimations of Na and K in the cat cerebral cortex (Bourke and Tower, 1966). The experimental levels of K were 94·5 µM. per g. and Na 57·4 µM. per g. The calculated levels of K and Na after allowing for the contamination of the extracellular fluids were K 166 µM. per g. and Na 18 µM. per g. The potassium level had been corrected by an increase of over 50 per cent and the sodium level had been decreased by over 60 per cent. Values for the sodium and potassium levels in nerve are given in *Table III*.

Table III.—The Approximate Concentration* of Sodium and Potassium in Nerve and Plasma

Subject	In Nerve		In Plasma	
	Na	K	Na	K
Carcinus (crab)	41	422	518	7·6
Loligo (squid)	50	400	440	22
Sepia (cuttlefish)	43	360	450	17
Cat cortex	18	166	135	5·0

* Concentrations are given in mEq./l.

Concentration and Activity

In gravimetric or flame photometric analysis the levels of sodium or potassium that are estimated are the total concentrations. There is the possibility that some of the potassium or sodium within the muscle or nerve could be bound to a substrate and not be freely available. This would mean that the effective concentration (activity) would be less than the total concentration.

One method of determining the activity of ions in solution is to use special ion-sensitive glass (Eisenman, Rudin, and Casby, 1957). Using intracellular electrodes of sodium- or potassium-sensitive glass it has been possible to determine the activity of ions within the cells. Hinke (1961) measured the activity of sodium and potassium using micro-electrodes inserted into squid axon. He also measured the activity in extruded squid axoplasm and calculated that 80–100 per cent of the potassium in the axon was free in solution but that only 76 per cent of the sodium was free.

Lev (1964) made micro-electrodes from ion-sensitive glass, the electrodes having a tip size of less than 1 µ. These electrodes were inserted into frog muscle-fibres. He found that more than 95 per cent of the potassium within the muscle-fibre was free but that less than 50 per cent of the sodium ions were free. It appeared that much of the sodium was bound or sequestered.

These measurements are important since it is the activity of the ions that will determine their effective concentration gradient and their role in establishing membrane potentials.

In a recent account of glass micro-electrodes Kostyuk, Sorokina, and Kholodova (1969) describe their measurements of the internal potassium and sodium concentration of frog muscle and snail neurons. In frog muscle the apparent activity coefficient for K was 0·73 and that for sodium was 0·39. The concentration of potassium as estimated by the electrode was 0·096 M whilst that determined by chemical analysis was 0·135 M. For sodium the value determined by the internal electrode was 0·0073 M and that by chemical analysis 0·0188 M. For *Helix* neurons the internal potassium as determined by the electrode was 0·073 M and that by chemical analysis 0·093 M. The values for internal sodium levels in the neuron were 0·013 M by the glass electrode and 0·031 M by chemical analysis.

Kostyuk and others (1969) suggested that there may be a decrease in the energy of the ions in protoplasm since normally one compares the activity of the ion in dilute solutions of KCl or NaCl and in protoplasm it is likely that there could be polyelectrolytes that would affect the activity coefficient. In model systems, where the sodium or potassium was diluted with polyelectrolytes such as polyacrylamide and acrylic acid, they found that the activity of the sodium changed with the concentration of the polymer. The diminution of the activity coefficient of sodium in the polyelectrolyte was probably caused by electrostatic inactivation in an intense field of closely placed charges. It is likely that similar field effects exist in living membranes and should be considered in studies of the roles of ions in living systems.

Model Systems

One way of explaining the manner in which cells could have a higher concentration of potassium ions inside than outside is by using model systems that invoke well-known physical and chemical systems. Some of these model systems will now be described since they indicate the type of answer that was used to explain the biological phenomenon. They are all answers 'of their time' and reflect the level of explanation that was current. They are most probably wrong or of very limited use in explaining the high internal K^+.

GIBBS-DONNAN EQUILIBRIUM

Cytoplasm contains a high proportion of proteins, peptides, and amino-acids which carry a charge. The cell behaves as if it is permeable to some ions but impermeable to other ions and to large charged particles; the large particles are too big to penetrate the small pores of the cell's membrane. The impermeable ions cannot, for some reason, get through the membrane. The large charged particles will affect the distribution of the permeable small ions and the theory concerning the distribution was developed by Gibbs (1906) and investigated by Donnan (1924).

If one considers that the proteins have a predominantly negative charge and assumes that ions such as K^+ and Cl^- are able to penetrate the membrane freely, then the protein will affect (a) the distribution of the permeable ions, (b) the osmotic pressure of the system, and (c) the electrical potential across the membranes.

If one considers a system where we initially have 100 K^+ and 100 Cl^- ions and a membrane equally permeable to K^+ and Cl^- ions, then there will be an equal concentration of K^+ and Cl^- ions either side of the membrane:—

In	Out
K_i^+ 50	K_o^+ 50
Cl_i^- 50	Cl_o^- 50

$$K_i \times Cl_i = K_o \times Cl_o : 50 \times 50 = 50 \times 50.$$

The products of the diffusible ions on both sides of the membrane are equal at equilibrium conditions: $K_i \times Cl_i = K_o \times Cl_o$.

Now let us add non-permeable large charged anions (proteins) to the inside. There will be a change in the distribution of the ions, especially the K^+ which will be attracted by the large negatively charged protein. The new distribution of ions may be as below:—

Inside	Outside
Protein$^-$	
K_i^+ 90	K_o^+ 10
Cl_i^- 10	Cl_o^- 90

$$K_i \times Cl_i = K_o \times Cl_o : 90 \times 10 = 10 \times 90.$$

Such a system would help explain how the cell has a very much higher concentration of potassium and a much lower concentration of chloride ions inside than outside. The equilibrium is called a Gibbs-Donnan equilibrium and it assumes here that the membrane is impermeable to sodium ions.

Diffusion of H^+ Ions

Hydrogen ions diffuse rapidly and there are biochemical systems within the cells for the rapid production of hydrogen ions.

A model where it is possible to use the high concentration of hydrogen ions to build up a gradient of potassium ions was developed by Netter (1928). The system consists of two compartments, A and B, separated by a collodion membrane (*Fig.* 3). In

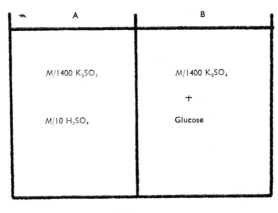

Fig. 3.—Model systems. Two compartments separated by a collodion membrane which will let the ions through. Initially the concentration of potassium ions in A and B are identical but after 13 days the concentration of potassium ions in A is twenty-five times that in B.

one compartment is placed a solution of $M/1400$ K_2SO_4 and $M/10$ H_2SO_4. In the other compartment is placed $M/1400$ K_2SO_4 together with glucose to balance the osmotic pressure across the membrane.

Initially the ratio of potassium ions in the two compartments A and B was 1 : 1. The ratio of H_A^+ : H_B^+ was 10^6 : 1.

The H^+ ion rapidly diffused from compartment A to compartment B following its concentration gradient. In doing this it tended to build up a positive charge in B and this was partially balanced by K^+ ions moving from B to A. At the end of 13 days when the concentrations were measured $K_A : K_B$ was 25 : 1. $H_A^+ : H_B^+$ was 52 : 1. There had been a concentration of potassium ions in A of twenty-five times.

If the system was left for a long time then $K_A^+ : K_B^+$ would become 1 : 1 and $H_A^+ : H_B^+$ would become 1 : 1.

The system demonstrates a method by which it is possible to get a twenty-fivefold concentration of potassium ions on one side of a permeable membrane compared to the other side. It could be similar to the method by which the cell accumulates potassium provided that the cell had a method of rapidly producing fast-diffusing hydrogen ions at one site within the cell and that these hydrogen ions could then diffuse out across the cell membrane.

DIFFERENTIAL MEMBRANE SOLUBILITY

Another model system which allowed the development of a concentration gradient was shown by Teorell (1933). He set up a tube containing sodium benzoate solution, carbon tetrachloride,

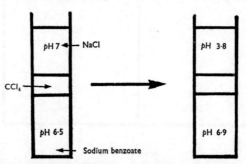

Fig. 4.—Model systems. Two compartments separated by carbon tetrachloride. Benzoic acid will dissolve in carbon tetrachloride and pass into the NaCl solution and so change the pH from 7 to 3·8.

and sodium chloride solution as shown in *Fig. 4*. The pH in the two compartments was 7 and 6·5.

Sodium ions and benzoic acid will dissolve in carbon tetrachloride but the hydrogen ion will not dissolve. The sodium benzoate reacts with water in the solution to form benzoic acid which can then dissociate:—

$$NaB + HOH = NaOH + HB; \quad HB = H^+ + B^-.$$

The undissociated HB will dissolve in the carbon tetrachloride and diffuse into the sodium chloride compartment where it dissociates to form H⁺ and B⁻. This increases the hydrogen ion concentration of the sodium chloride solution to a pH of 3·8.

The model indicates how a semipermeable solution (= membrane) can affect the movement of ions and so help to build up a concentration gradient. There is an increase in concentration of hydrogen ions in the sodium chloride solution by several hundred times. If this model was linked to a collodion membrane (model of the type shown in *Fig.* 3) so that the sodium chloride solution was free to diffuse it could lead to a change in the sodium concentration of the solution as the hydrogen ions diffused rapidly across the collodion membrane.

ACCUMULATION OF CHLORIDE IONS

In this model there are two compartments, A and B. One compartment is so large a volume that small changes in the concentration of ions can be neglected. The compartments are separated by a semipermeable membrane. In the small compartment A is 0·16 N NaCl solution. In the large compartment is a solution of 0·16 N HCl (*Fig.* 5). The hydrogen ions diffuse rapidly from compartment B to compartment A and in doing so they set up a potential gradient. This is then compensated by a movement of chloride ions (though there may be a small movement of sodium ions too).

After 1 day the concentration of chloride ions in the two compartments was such that the chloride concentration in A was 0·19 N whilst that in B was effectively the same as initially (0·16 N) though of course there was a very small fall in the chloride level in B. This depends on B being so large that the chloride loss was not detectable.

In all these models certain points in common should be noted.

1. They can illustrate how it is possible to develop a concentration gradient of a permeable ion across a membrane.

2. Often the system is 'open-ended', i.e., it does not consider the system at final equilibrium conditions where the levels of ions on both sides of the membrane would be equal concentrations.

3. They usually involve an initial change in the concentration gradient of some other ion such as the hydrogen ion.

4. The Gibbs-Donnan equilibrium is a more stable one though there is also the osmotic balance to be considered and the ionic imbalance may be affected by the energy change put in to compensate for the osmotic and potential imbalance.

The advantage of all these systems is that they indicate how *non-living* models can build up concentration gradients of ions similar to those that exist in living cells. The models do not involve any vital force. They sometimes beg the question by starting off with a high concentration of an ion such as hydrogen ions (much as the recipe for becoming a multimillionaire starts 'take one million pounds and invest it'—you have to get the first million to begin). One has to develop some chemical method for the rapid production of hydrogen ions.

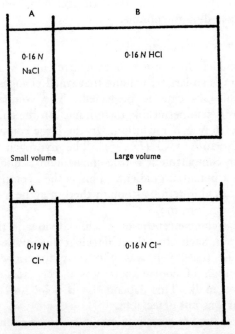

Fig. 5.—Model systems. Two compartments separated by a collodion membrane. A is much smaller than B. Chloride ions move from B to A and raise the concentration in A above that in B even though initially the chloride concentration in A = B.

The reason for discussing these systems is not that they provide the correct answer as to why there is a high concentration of potassium ions and a low concentration of chloride ions inside the cell. They indicate the type of answer that was accepted by research workers at the time. They used physicochemical models of the time to explain how a living system could work. In fact they were all explanations of their time.

The Sodium Pump

Höber (1945) pointed out that the idea that the cell membrane could have a selective permeability to ions such as potassium and sodium was at first considered to be quite fantastic, but experiments on the red blood-cell membrane quickly established the facts of anion permeability. It was then thought that there could be a selectivity on the basis of lipid solubility across the membrane or the size of the molecules, these ideas being based on simple physicochemical models.

However, evidence gradually accumulated that some other idea might be necessary in order to explain the ionic composition within cells since differential membrane permeability did not seem to be enough.

The new idea was that of the sodium pump.

Sodium Pump in Muscle

Boyle and Conway (1941) thought that the muscle-cell membrane was permeable to potassium ions and to chloride ions but *not* to sodium ions. They thought that the uptake of potassium ions was a natural consequence of the increase of organic phosphates within the cell and that these large anions would act through a Gibbs-Donnan equilibrium and trap the potassium ions. The views of Boyle and Conway with regard to the membrane permeability were supported by many people up to 1952.

Even at the time of Boyle and Conway's writing there was quite a lot of evidence that the muscle membranes were permeable to sodium.

Fenn and Cobb (1934) showed that rat muscle lost potassium and gained sodium during activity and that when the muscle was relaxed there was an increase in the intracellular potassium and a loss in the intracellular sodium. Heppel (1939, 1940) raised rats on a diet deficient in potassium and found that almost half of the potassium in the muscles had been replaced by sodium. The sodium was freely diffusible. If the rats were then given a diet containing potassium, this was taken up into the muscles. Steinbach (1940) showed that isolated frog muscles soaked in potassium-free Ringer solution exchanged 40 per cent of their internal potassium for sodium. On replacing the muscles in normal potassium-containing Ringer solution the muscles accumulated potassium and ejected the sodium.

There was thus a reasonable amount of indirect evidence that the muscle membrane was permeable to sodium ions and this was confirmed by direct experiments using labelled sodium. Heppel (1940) showed that labelled sodium could penetrate the rat muscle membrane at a rate equal to, or even faster than, potassium.

Dean (1941) suggested that the muscle membrane was permeable to both sodium and potassium ions: 'There must be some sort of pump, possibly localized in the fiber membrane, which can pump out the sodium, or, what is equivalent, pump in the potassium.'

The system pumped the ions against the concentration gradient. It required energy to do this and when the aerobic or anaerobic respiration of the muscle was stopped the sodium pump stopped and the sodium concentration within the muscle increased. Thus it was possible to load muscle-cells with sodium by storing them at low temperature in Ringer lacking potassium. Since active processes are inhibited at low temperatures the sodium was accumulated and the potassium was lost. If the muscle was now transferred back to Ringer containing normal potassium concentration at room temperature the sodium was pumped out and the potassium accumulated. There was a link between the sodium pumped out and the potassium pumped in.

SODIUM PUMP IN NERVE

Analyses by Keynes and Lewis (1951) of the sodium and potassium levels in nerve were followed by experiments by Hodgkin and Keynes (1955) on the ionic fluxes across the nerve membrane. It became clear that the nerves could take up potassium ions and pump out sodium ions. Further experiments by Caldwell and Keynes (1959, 1960) and Caldwell, Hodgkin, Keynes, and Shaw (1960a, b) showed that the sodium efflux from squid axon was reduced by treatment with cyanide, azide, or dinitrophenol and that the sodium efflux in these poisoned nerves could be restored by injecting ATP or selected phosphagens into the axon (*Fig.* 6). Baker (1965) showed that between 2·7 and 4 sodium ions were pumped out of nerve for every ATP broken down.

Nerve thus contains an active sodium–potassium pump in the membrane. The pump moves the ions against the concentration gradient and requires energy to do so. The linkage between the sodium and potassium ions is approximately 3 Na ions pumped out for every 2 K ions pumped in.

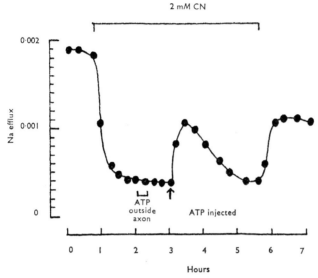

Fig. 6.—The sodium efflux from squid axon. Labelled sodium ions are loaded into the squid axon, then they are pumped out. The pump is stopped when the nerve is placed in 2 mM cyanide solution and is not affected by application of ATP outside the axon. Injecting ATP inside the axon allows the sodium efflux to start again. (*Reproduced from* Caldwell, P. C., Hodgkin, A. L., Keynes, R. D., *and* Shaw, T. I. (1960a), *J. Physiol., Lond.*, **152**, 561–590.)

SODIUM PUMP IN RED BLOOD-CELLS

It was considered by some authors even as late as 1940 that the red blood-cell was impermeable to Na and K. The level of K and Na in the red blood-cell was explained as a speciality of the nucleated precursor of the red blood-cell and that the cell, on losing its nucleus, became impermeable to Na and K ! This was in spite of the evidence from many laboratories which indicated that Na and K could penetrate the red blood-cell membrane. But the power of contemporary opinion was such that these experiments were either ignored or explained away.

Accurate analysis of the red blood-cells showed that the Gibbs-Donnan equilibrium was insufficient to allow for the established concentration gradient of K and Na across the membrane. Maizels and Patterson (1940) showed that human red blood-cell could take up sodium ions when stored *in vitro* and then expel the sodium ions if the cells were placed *in vivo*. Harris (1941) showed that the red blood-cells would lose potassium when they were stored in the cold. On rewarming the cells accumulated potassium provided that glucose or pyruvate was present. The pumping in of potassium was inhibited by fluoride. The work of several groups of workers, such as those of Maizels (summarized 1954), Post and Jolly (1957), Glynn (summarized 1957a, b), and Whittam (summarized 1964), has clearly shown the basic properties of the sodium–potassium pumping mechanism in the red blood-cell membrane.

In summary, the sodium–potassium pump is a new feature that is different from the previous physicochemical models in that it uses energy to pump ions against the concentration gradient. It is *par excellence* an example of a vital force, a force that exists in living systems which, even today (1971), we do not understand sufficiently to write out all the chemical equations underlying it. Nevertheless, even though it is a vital force, it is still, we assume, one that is ultimately explicable in terms of the chemistry and physics of the next few decades.

LINKAGE BETWEEN THE SODIUM AND POTASSIUM PUMP

Harris and Maizels (1951) found that there was a linkage between the sodium pumped out of the cell and the potassium pumped in, since there was a marked reduction in sodium efflux from red blood-cells if they were placed in solutions containing less than 1 mM external potassium concentration.

Further studies by Whittam and his colleagues, Glynn, and Post and Jolly showed that there was a clear linkage between the sodium efflux and the potassium influx. The ratio observed varied according to the conditions and the following ratios have been found:—

Na : K	Author
2 : 1	Harris, 1954
1 : 1	Glynn, 1956
3 : 2	Post and Jolly, 1957
1 : 1	McConaghey and Maizels, 1962

It is possible that the ratio could vary according to the experimental conditions as well as natural conditions.

The reconstituted ghost technique of Hoffman and others (1960) and Glynn (1962) showed that the sodium efflux was most effective when the sodium ions were on the inside of the red blood-cell membrane and the potassium ions on the outside. The pump was therefore orientated across the membrane so that the sodium would attach more easily on one side and the potassium on the other.

ATPase

Skou (1957, 1960) extracted an ATPase from crab nerve. This ATPase had special properties in that it required specific ions for maximal activity. Its activity was greatly increased in the presence of Na, K, and Mg. The presence of Na alone did not stimulate the ATPase, nor did the presence of K alone. Instead it required both Na and K for maximal activity. *Fig.* 7 shows a model system of the manner in which the Na–K ATPase could change

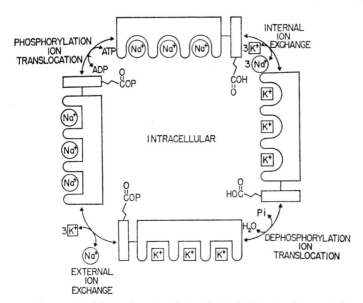

Fig. 7.—Model for the role of the Na–K ATPase in the coupled transport of Na and K across the membrane. The form of the ATPase changes following the dephosphorylation so that the molecule changes its affinity for K and increases that for Na. (*Reproduced from Hokin, L.* (1969), '*J. gen. Physiol.*', **54**, 327s–342s.)

its shape and affinity for Na or K and hence help in the passage of the material across the membrane. Energy is involved in the change of the configuration of the ATPase molecule (Hokin, 1969). Though this model is not complete it does provide some indication of the type of reaction that could be happening.

A similar Na–K–Mg ATPase has been found in red blood-cell membranes, crab axon, mammalian muscle membrane, and mammalian nerve-cells and it is considered to play an important role in the functioning of the Na–K pump. The enzyme is inhibited by cardiac glycosides. Other ATPases have been extracted from mammalian tissues but these differ in their properties. Three ATPases have been described: Na–K–Mg ATPase (as above), Mg ATPase, and Ca ATPase. Only the Na–K–Mg ATPase is inhibited by cardiac glycosides. It is also inhibited by parachlormercuribenzoate, calcium, cysteine, octylguanidine, oligomycin, and NEM (n-ethylmaleimide) (Bonting, 1970).

The ATPase from nerve synaptosomes has been added to an artificial membrane and the whole system has been shown to pump sodium and potassium and also to generate a potential (Jain, Strickholm, and Cordes, 1969).

Cardiac Glycosides

Compounds such as digitalis have been known since the time of William Witheringham to have a beneficial effect on the failing heart-beat. When the heart shows auricular fibrillation, addition of digitalis or a cardiac glucoside makes the heart beat more regularly, slowly, and strongly. Schatzman (1953) showed that the sodium pump in the red blood-cell was inhibited by low concentrations (10^{-8} g. per ml.) of glycosides such as strophanthin. This was confirmed by later workers who showed that a range of compounds such as scillaren A, digitonin, emicymarin, or ouabain (*Fig.* 8) would all inhibit the sodium pump. At low concentrations they had no effect on cellular respiration but instead acted selectively on the sodium–potassium pump. Glynn (1957b) studied the relative efficacy of different glycosides and taking scillaren A as 1, the relative effectiveness was emicymarin 2/5, digoxin 1/7, hexohydroscillaren 1/100, and alloemicymarin 1/300.

Injection of high concentrations of ouabain into squid axons by Caldwell and Keynes (1959) did not inhibit the sodium pump, but the application of even low concentrations to the outside of the axon did inhibit the sodium pump. This suggests that the site of the glycoside action may be on the outside of the membrane and that possibly the glycoside could act by competing with potassium for the active site. Glynn (1957b) calculated that as few as 1000 molecules of glycoside per red blood-cell would completely inhibit the sodium pump.

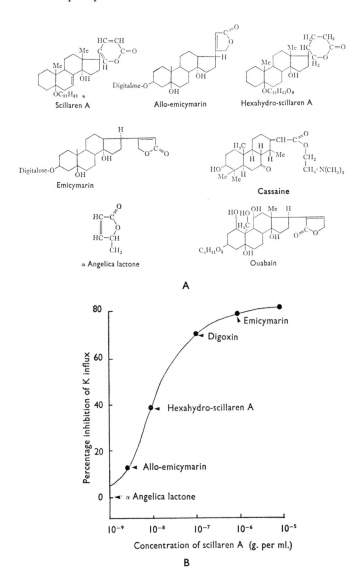

Fig. 8.—A, Formulae of the glycosides that can inhibit the sodium pump in red blood-cells. B, The effective concentration relative to scillaren A is indicated. (*Reproduced from Baker, P. F.* (1966), *Endeavour*, **96**, 166–169. *After Glynn, I. M.*)

The cardiac glycosides have become an important biochemical and physiological tool. When they have an effect at a low concentration it can usually be taken to indicate that the glycosides have interfered with the Na–K pump and the resulting condition is a result of this interference. In general the conclusion seems to be a valid one though it should be remembered that there is no such thing as an absolutely specific inhibitor.

Ouabain at higher concentrations can have additional effects. It can interact with and mobilize tissue calcium. It can stimulate mitochondrial respiration. These effects normally occur at higher concentrations than those used to inhibit the Na–K–Mg ATPase. Therefore, caution should be used in interpreting results where levels above 10^{-4} g. per ml. of ouabain or other cardiac glycosides are used.

Membrane Potential in Nerve and Muscle

Between the inside and the outside of a nerve-cell or a muscle-cell there is a steady electrical potential. This is called the 'membrane potential'. The membrane potential of the cell at rest is called the 'resting potential'.

Bernstein's Theory

The existence of the resting potential had been known for many years. Bernstein (1902) suggested that the membrane potential could be explained in terms of the selective permeability of the cell membrane to potassium ions. The cell would have a high concentration of potassium ions inside the cell and a low concentration outside the cell. The selective permeability of the membrane to potassium would establish a concentration (equilibrium) potential across the membrane.

Michaelis Collodion Membrane

The Bernstein theory was helped by the use of a model system. Michaelis (1928) developed a collodion membrane that had pores which would allow the passage of cations. The permeability to cations increased as the membrane dried out. If the membrane was placed between two different concentrations of KCl solutions there was a potential across the faces of the membrane with the positive pole directed towards the lower concentration of KCl. If the concentration of KCl on one side was ten times more than that on the other side of the membrane, then the potential across the membrane was found to be 58 mV. This is described by the Nernst equation which states that the potential is proportional to the concentration gradient of ions:—

$$E = \frac{RT}{nF} \log_e \frac{10}{1} = 58 \text{ mV},$$

where E is the potential in mV.; R, the gas constant; T, the temperature °Kelvin; n, the valency of the ions; F, Faraday.

The equation can be written $E = RT/nF \log_e c/c'$, where c and c' are the concentrations of the ions either side of the membrane.

Bernstein measured the potential between a live piece of nerve and the end of the nerve killed in KCl. This demarcation potential came to 35 mV. Similarly the measured potential between the intact muscle and the KCl-treated end of the muscle was 50 mV. The KCl-treated muscle had effectively the same potential as the inside of the muscle-fibre, so the demarcation potential measured the potential between the inside and the outside of the fibre. The value was low due to short-circuiting between the electrodes.

MUSCLE-MEMBRANE POTENTIAL

With the advent of glass capillary micro-electrodes Graham and Gerard (1946) were able to measure the membrane potential of skeletal muscle-fibres accurately without any short-circuiting (such as was present in the demarcation method used by Bernstein). The value of the muscle potential came to −98 mV. The value calculated from the Nernst equation, assuming that potassium was the responsible ion, was −100 mV.

NERVE-MEMBRANE POTENTIAL

The use of the internal electrode in the squid giant axon by Hodgkin and Huxley (1945) gave a value for the membrane potential of −45 mV. The corrected value allowing for junction potentials was −62 mV. The value calculated from the Nernst equation was −89 mV.

EFFECT OF CHANGING THE EXTERNAL POTASSIUM CONCENTRATION

The membrane potential was most affected by changing the concentration of external potassium ions, compared with changes in the external sodium or chloride concentration. Over a range of high external potassium concentrations the membrane potential obeyed the Nernst equation for potassium and had a slope of

58 mV. When the external potassium ion concentration was equal to the internal potassium concentration the membrane potential was zero. When K_o was greater than K_i the membrane potential became reversed and the inside was positive with respect to the outside of the membrane.

Over the physiological range, however, the membrane potential does not obey the Nernst equation. For changes of external potassium concentration below 5 mM there is relatively little change in the membrane potential. Hodgkin and Horowicz (1960) explained this in terms of an increase in the permeability of the membrane to sodium at these low K_o values. The equation that they used was

$$V = \frac{58 \log K_o + 0.001 Na_o}{140},$$

where 140 is the K_i value. *Fig.* 9 shows the theoretical curve from the Nernst equation and the experimentally obtained curve. The

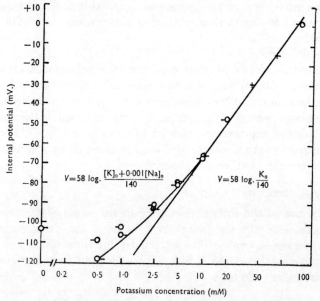

Fig. 9.—The change in resting potential of a single muscle-fibre with change in external potassium concentration. At low concentration below 5 mM K, which is the physiological range, the experimental observed resting potential deviates from the Nernst equation for potassium alone. It fits in with an equation where there is a limited permeability of membrane to sodium ions as indicated by the equation. (*Reproduced from* Hodgkin, A. L., *and* Horowicz, P. (1959), *J. Physiol., Lond.,* **148**, 127–160.)

latter fits the equation given above, allowing for an increased permeability to sodium ions. In snail neurons, if the external sodium concentration is lowered when the external potassium is lowered, then the potential fits in with the Nernst equation for potassium, indicating that there is an increased permeability to sodium ions at low K_o.

NATURE OF THE MEMBRANE POTENTIAL—EQUILIBRIUM POTENTIAL OR ELECTROGENIC ?

The membrane potential in nerve and muscle is thus explicable in terms of the concentration gradient of potassium ions between the inside and the outside of the membrane. The membrane behaves as if it were mainly permeable to potassium ions and impermeable to sodium or chloride ions. Hodgkin and Keynes (1955), in their study of the action of metabolic inhibitors such as cyanide, dinitrophenol, and azide on the sodium pump in *Sepia* axon, found that these metabolic inhibitors reduced the sodium flux from the axon (*Fig.* 10). There was some indication that dinitrophenol could affect the membrane potential. *Fig.* 11 shows the effect of adding 0·2 mM dinitrophenol to the squid axon. There was a slight decrease in the membrane potential (shown by the line 2 in *Fig.* 11). The membrane potential had been −65 mV. and the addition of dinitrophenol reduced the membrane potential by 3 mV. to 62 mV. They thought that this change was small and close to the experimental error.

In considering whether the sodium pump was electrogenic (i.e., produced a potential) they suggested that if there were 3 Na ions pumped out for every 2 K pumped in, then this would bring about a change of 1·8 mV. in the membrane potential. The point is fully discussed by Hodgkin and Keynes (1955) in the following passage:—

'A coupled system which absorbed one K$^+$ for each Na$^+$ ejected would be electrically neutral in the sense that it would transfer no charge across the membrane. With such a mechanism one would expect changes in the activity of the pump to have no immediate effect on membrane potential, while alterations in membrane potential ought not to have much effect on the activity of the pump. Both predictions agree with our observations, but this is not really good evidence that the pump is neutral. If a chemical reaction involved in the sodium pump gave 10,000 cal. per mole (as in hydrolysis of ATP), it would be capable of driving

30   THE ELECTROGENIC SODIUM PUMP

ions against an electrochemical potential difference of 430 mV. In this case the change of 20 mV. produced by applied currents might reduce the efflux of sodium by an insignificant amount. A reservation must also be made about the lack of any change in membrane potential when the sodium pump is blocked. The

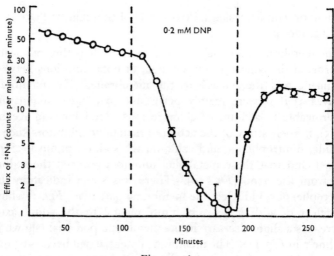

Fig. 10.—A

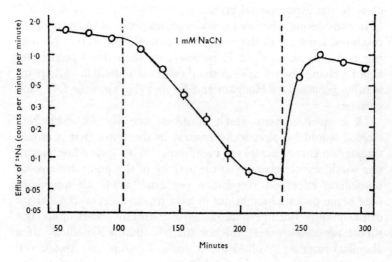

Fig. 10.—B

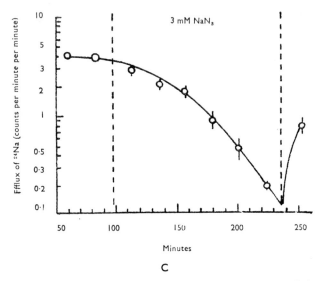

Fig. 10.—The effect of metabolic inhibitors such as 2,4-dinitrophenol, NaCN, or NaN_3 on the efflux of sodium from squid nerve. They all reduce the efflux, i.e., inhibit the sodium pump. (*Figs.* 10, 11 *reproduced from* Hodgkin, A. L., *and* Keynes, R. D. (1955), *J. Physiol., Lond.*, **128**, 28–60.)

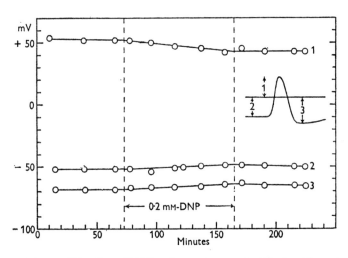

Fig. 11.—The effect of DNP on the membrane potential of squid axon. **1**, Height of action potential. **2**, Resting potential. **3**, Positive phase. DNP does bring about a slight decrease in the resting potential.

argument is clearest for the squid axon, which has a resting membrane resistance of only about 1000 Ω per sq. cm. The sodium efflux from the squid axon is about 50 pmole per sq. cm. per second, and it may be assumed that, as in *Sepia* axons, two-thirds of this is balanced by an active uptake of potassium. If the remaining one-third consists of a stream of sodium ejected as ions, the pump would generate a current of 1·8 µA., and its direct contribution to the membrane potential would be only 1·8 mV. The lack of any marked change in resting potential on applying DNP to the squid axon is therefore not conclusive evidence that the pump is neutral.

'Although we have presented evidence for a loose or partial coupling between the secretory movements of sodium and potassium, it seems that there cannot be a rigid linkage between sodium efflux and potassium influx. Removing the potassium from the external medium reduces the sodium efflux by a factor of about 3, but much greater reductions are observed at low temperatures or under the influence of metabolic inhibitors. The residual sodium efflux into potassium-free solutions is not an exchange for external sodium, since it is increased by substituting choline or dextrose for sodium. When sodium ions are extruded into a salt-free solution they must be accompanied by anions, but there is no indication as to the nature of these substances, and we do not know whether they leave the nerve passively or are actively extruded with sodium' (Hodgkin and Keynes, 1955).

Electroneutral Pump versus Electrogenic Pump

It is useful to distinguish between the electroneutral sodium pump and the electrogenic sodium pump.

In the electroneutral sodium pump the sodium ions are pumped out and the potassium ions are pumped in. If a number of sodium ions are pumped out equal to the number of potassium ions pumped in, i.e., the pump is 1 Na to 1 K, then the pump would be electroneutral. Or if there was an anion moving out with the sodium ions, then the pump could be 3 Na to 2 K with the anion balancing the charge, and still be electroneutral.

In an electrogenic sodium pump the number of cations moving out of the membrane is greater than the number of cations moving in and there is no anion movement which would balance the charge.

THE NATURE OF SCIENTIFIC PROGRESS 33

This unbalanced ionic movement would produce a current, but for the potential to develop across the membrane, it is necessary that the membrane resistance is high and that there is little or no short-circuiting of the current by movements of other ions such as chloride ions.

If the experimental conditions can be set up so that the pump is maximally stimulated and the membrane resistance is high and there is no short-circuiting of the current, then the sodium pump can develop a potential of up to 30 mV. across the membrane.

An alternative method of developing the potential would be for the metabolic processes of the ATPase–Na–K pump to have an electron transfer system involved and the movement of the electrons could under some circumstances develop a potential. This type of electrogenic potential, produced by electron or proton movements, will not be considered in the present account.

We can make the differences between the two views concerning the nature of the potential across the membrane more clear by oversimplifying and proposing the following two schemes, which we shall call the classic view and the electrogenic sodium pump.

CLASSIC VIEW

1. The metabolic activity of the cell pumps sodium out and potassium in.
2. This builds up a high concentration of potassium inside the cell.
3. The membrane potential is due to this imbalance of potassium ions across the membrane.
4. If the pump is switched off there is a very slow fall in the internal potassium concentration and a small and slow decrease in the membrane potential.
5. Stimulation of the sodium pump by injecting sodium ions into the cell increases the amount of sodium pumped out and also increases the amount of potassium pumped in. This would lead to an increase in the membrane potential (E_K) and the cell hyperpolarizes.
6. The increase in membrane potential following stimulation of the sodium pump is not due to the electrogenic nature of the pump but is due to the uptake of potassium around the cell and

hence a decrease in K_o and an increase in E_K. This scheme is shown diagrammatically in *Fig.* 12.

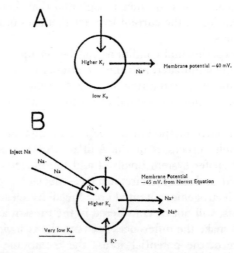

Fig. 12.—The classic view of the increase in membrane potential during the activity of the sodium pump. It is due to a decrease in the external potassium ions concentration and hence an increase in E_K. A, An explanation of the increase in membrane potential after injection of Na. The membrane potential is determined by K_o/K_i. B, When Na is injected it is pumped out. K^+ is pumped in to balance. This reduces K_o to a very low value and increases K_i. But the amount by which it can increase by this method is only 2–3 mV. The observed value can be up to 30 mV. Thus this explanation is not a valid one.

SUMMARY OF THE CLASSIC VIEW CONCERNING THE NATURE OF THE RESTING POTENTIAL IN NERVE AND MUSCLE

1. Chemical analysis shows that the cells have a higher internal potassium ion concentration than is present in plasma.

2. Early measurements of the potential between the outside of a nerve and the killed end of the nerve (demarcation potential) indicated that the membrane potential was about 25–30 mV.

3. Improved chemical analysis of the potassium content of nerve indicated that the membrane potential should be approximately 60 mV. if it was determined by the Nernst equation with potassium as the main ion. (Note that the membrane potential is consistent for a given tissue or cell but that it will vary between tissues such as frog muscle, heart muscle, or squid nerve.)

4. Improved techniques for the determination of the membrane potential for cuttlefish giant axon gave a value of 48 mV. Allowing

THE NATURE OF SCIENTIFIC PROGRESS 35

for junction potentials the value came to 62 mV. The theoretical value for E_K was 89 mV.

5. For frog muscle the predicted value from the potassium concentration of the cells and the plasma came to 100 mV. The experimentally observed value was 97·5 mV.

6. For frog muscle and squid nerve it appears that the experimentally determined membrane potential is below the potassium equilibrium potential (E_K). Thus the ionic theory is more than sufficient to explain the membrane potential of nerve and muscle.

7. The membrane potential can reach the potassium equilibrium potential (E_K) if the permeability to potassium is increased. Thus in heart muscle the action of acetylcholine is to increase the permeability of the muscle membrane to potassium ions so that the muscle potential hyperpolarizes and approaches the value of E_K.

8. There is also a significant permeability to sodium across the resting muscle and nerve membrane and this brings the membrane potential below that predicted by E_K.

9. It is a more economical hypothesis to consider that the membrane potential is fully explained in terms of the potassium equilibrium potential than to invoke additional phenomena.

10. The sodium pump in squid axon, if stopped, only brings about a fall of 3 mV. in the membrane potential. Nerves or muscles can be poisoned with dinitrophenol or cyanide and show a membrane potential of 60 mV. for several hours.

11. There would not seem to be any metabolic component directly adding to or acting on the resting membrane potential.

ELECTROGENIC SODIUM PUMP

This view is that in some nerve-cells and muscle-cells there is a significant part of the membrane potential contributed by the activity of the electrogenic sodium pump. This portion of the membrane potential is important because it can contribute to the difference in the threshold level that can then determine whether a nerve- or a muscle-cell will give an action potential or not. The view concerning the electrogenic potential is summarized as follows:—

1. The metabolic activity of the cell pumps sodium out and potassium in.

2. This builds up a high concentration of potassium within the cell.

3. The resting membrane potential is due (50–99 per cent) to the imbalance of potassium ions across the membrane.

4. When the sodium pump is switched off there is a drop in the membrane potential *at once*. This may be a small decrease, when the electrogenic sodium pump makes a small contribution, or up to a 50 per cent decrease in potential when the electrogenic sodium pump makes a bigger contribution to the membrane potential.

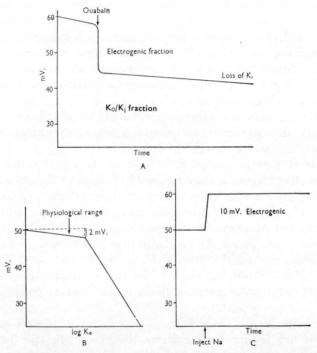

Fig. 13.—The electrogenic sodium pump explanation. **A**, Some nerves have a membrane potential which is decreased by up to 10 or 20 mV. on addition of ouabain. This fraction of the membrane potential is electrogenic. The remainder of the membrane potential is due to E_K. **B**, If the external potassium concentration is lowered there is only an increase of 2 or 3 mV. in the membrane potential (more often there is a decrease in the membrane potential). **C**, If sodium is injected into the neuron there is an increase in the membrane potential by more than 10 mV. This cannot be explained in terms of reduction of K_0 and an increase in E_K.

In addition, if the metabolic system is inhibited for a long time, there will be a slow decrease in the internal potassium ions concentration and this will bring a further slow decrease in the membrane potential (*Fig.* 13).

5. Stimulation of the sodium pump by injecting sodium ions into the cell increases the amount of sodium pumped out of the cell. It also increases the amount of potassium pumped in.

The fall in the potassium concentration immediately around the cell membrane is not the significant factor bringing about the hyperpolarization of the cell (*Fig.* 13 B), since the cell is on the part of the curve where there is very little change in membrane potential for a reduction in the external potassium concentration (*Fig.* 13). Thus decreasing the potassium concentration around the cell from 4 mM to 0 mM may bring about an increase in the membrane potential of the cell by 2 mV. Injection of sodium ions and stimulation of the pump can bring about an increase in the potential of more than 30 mV.

For these reasons it is suggested that many cells may have the sodium ions entering the cell at such a rate that the electrogenic sodium pump may make a considerable contribution to the membrane potential.

The pumping of the sodium ions out of the cell leads to the hyperpolarization of the cell. This is because the sodium pumped out is not balanced by the potassium coming in. The pump is electrogenic—it generates a potential.

Why was the Contribution of the Electrogenic Pump to the Resting Potential Missed?

The study of the resting potential of nerve- and muscle-fibres showed a good correlation between the experimental values and those predicted by the Nernst equation, assuming that potassium was the only ion involved in the system. If anything, the E_K was greater than the membrane potential and so there was more than sufficient force present across the membrane to explain the membrane potential.

The electrogenic sodium pump had been considered by Hodgkin and Keynes (1955), who found that the calculated value for the electrogenic sodium pump, assuming that there was an imbalance of 3 Na to 2 K, was only 1·8 mV. This would not be a very significant contribution to a resting potential of 62 mV. Their experimentally determined value was 2–3 mV., which fitted in with the theoretical value. Grundfest (1955) carried out a series of experiments on the effects of injecting sodium ions into squid axons and, though the effects he obtained from these experiments were not very great, he considered that the electrogenic sodium pump might play an important role in establishing nerve potentials. However, his experimental basis for this did not seem strong enough to balance the results obtained by Hodgkin and Keynes. Hodgkin (1958) considered that there was almost an equal amount of evidence that the sodium potassium exchange might be 1 : 1, in which case there would be no argument for an electrogenic component.

The problems are:—
1. What is the evidence that the sodium pump is electrogenic?
2. What contribution can the sodium pump play towards the establishment of the membrane potential of nerve and muscle?
3. Can the sodium pump and the metabolic activity of the nerve- or muscle-cell vary significantly and change the electrical activity and responses of the nerve and muscle?

The answers to these questions will be given in the next section.

ELECTROGENIC SODIUM PUMP IN NERVE-CELLS AND AXONS

As we have seen, there is no dispute that there is a mechanism for pumping sodium ions out of nerve- and muscle-cells and for pumping potassium ions in. The question is whether the pump is electroneutral, the sodium pumped out balancing the potassium pumped in, or electrogenic, more sodium being pumped out than potassium pumped in. Other ions could also play a role but in this account they will generally be ignored.

An easy way to stimulate the sodium pump is to inject sodium ions into the cell. Calculation based on the labelled sodium efflux from cuttlefish axon indicated that if 3 Na were exchanged for 2 K the calculated change across the membrane would be 1·8 mV. Adding metabolic inhibitors such as dinitrophenol to the *Sepia* axon had only a small direct effect on the membrane potential. The maximum immediate fall was 3 mV.

Thus the electrogenic sodium pump seemed to play little or no role in establishing the 62 mV. resting potential across the *Sepia* axon and it was considered that the pump might have the ratio 1 Na : 1 K (Hodgkin, 1958).

POST-TETANIC HYPERPOLARIZATION (PTH)

If a nerve-trunk is rapidly stimulated sodium ions enter the nerve during the action potentials. Such a burst of tetanic stimulation often leads to a hyperpolarization of the nerve and this is called post-tetanic hyperpolarization (PTH) (*Fig.* 14).

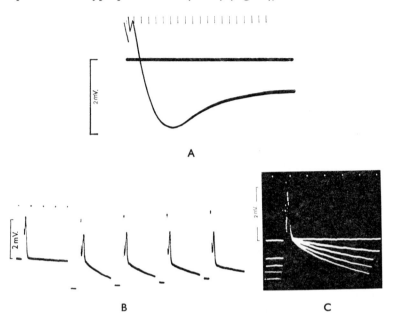

Fig. 14.—Post-tetanic hyperpolarization (PTH). A, Record taken after the nerve has been given a tetanic burst of stimulation. There is a marked downward hyperpolarization of the nerve potential. B, The record at the extreme left was taken before, and the rest taken 1·7, 5·1, 8·5, and 15·3 seconds after, the end of a 10-second period of repetitive stimulation. Note the increase in the downward displacement (PTH) of the record. C, Records as in B but superimposed. (*Reproduced from Ritchie, J. M., and Straub, R. W.* (1956), '*J. Physiol., Lond.*', **134**, 698–711.)

Ritchie and Straub (1957) found that the PTH in non-medullated small nerve-fibres of the rabbit cervical sympathetic nerve-trunk was up to 4 mV. in size. The PTH disappeared if the nerve was soaked in dinitrophenol or ouabain. The nerve recovered if it was washed with fresh Locke solution and would again show PTH. Ritchie and Straub thought that the PTH was due to the rapid uptake of potassium ions by the nerve-cells and hence a lowering of the K_0 around the nerve-cell. This would raise the E_K and hence lead to the observed hyperpolarization.

Connelly (1959) extended the experiments on PTH using frog nerve-trunk. He suggested that the development of the PTH was, in general, parallel to the rate of sodium excretion at the nodes. There was also an increased oxygen consumption following PTH and he suggested that there could be a metabolic basis for the PTH. The time taken for the nerve to recover from PTH in high potassium medium was less than in normal potassium Ringer. More time was required to recover in low external potassium Ringer. This finding contradicted the view that the PTH was dependent on the nerve reducing the external potassium concentration and so raising E_K. Connelly concluded that the PTH was most likely due to an electrogenic sodium pump in the nerve-trunk.

Later work by Straub (1961) and Greengard and Straub (1962) showed that the PTH was blocked by antimycin A, which inhibits the activity of cytochrome-c reductase. Other metabolic inhibitors such as azide, cyanide, dinitrophenol, or iodoacetate reduced or blocked the PTH. Removal of glucose from the surrounding solution would also reduce the PTH. One important point was that the PTH could be greater than that predicted from E_K and so could not be due to the removal of the potassium around the neuron. The experiments showed the manner in which the metabolic activity of the cell brought about a potential change across the membrane. Straub thought that PTH was best interpreted in terms of the activity of an electrogenic sodium pump.

The PTH was in general quite small and required a considerable burst of electrical stimulation to elicit it. The measurements of the PTH were also technically difficult since they were usually made by means of the sucrose-gap technique and this method can develop considerable junction potentials which might require a large correction. However, in the experiments that have just been described, the authors were working on the differences in the potential of the nerve before and after the tetanus and hence the junction potential values would have cancelled each other out.

In 1968 Rang and Ritchie working on non-medullated mammalian nerve-fibres realized that the PTH was small because much of the current was being short-circuited by diffusible ions such as chloride ions. They replaced the chloride ions in their Locke solution by the larger non-diffusible isethionate ion. They found that the PTH could now be up to 35 mV. in size. This was very much larger than could be explained in terms of E_K and was only explicable in terms of an electrogenic sodium pump.

Snail Neurons

Kerkut and Thomas (1965) were able to get larger potential changes in snail neurons by injecting sodium ions into selected cells. The large neurons of the snail brain lie peripherally in the ganglia and so are easy to observe and it is possible to insert several electrodes into one neuron and to remove or change the surrounding solution (Kerkut, 1967). If the membrane potential was measured by means of a potassium-acetate-filled microelectrode and then a low-resistance sodium-acetate-filled electrode was inserted into the same cell as the sodium diffused from the electrode into the cell, there was a marked hyperpolarization of the cell (*Fig.* 15). The membrane could hyperpolarize by 30 mV.

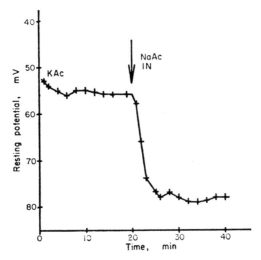

Fig. 15.—Injection of sodium ions into snail neurons brings about a marked hyperpolarization of up to 30 mV. in 10 minutes. (*Figs.* 15, 16, *reproduced from Kerkut, G. A., and Thomas, R. C.* (1965), '*Comp. Biochem. Physiol.*', **14**, 167–183.)

in 10 minutes. The hyperpolarization could be reduced by the addition of ouabain (*Fig.* 16) or by the reduction of the external potassium ion concentration.

This hyperpolarization was often much larger than that found by workers on PTH. The potential could not be explained in terms of reducing the external potassium concentration around the neuron (increasing E_K) since it was larger than that predicted from the Nernst equation. Furthermore, in these neurons,

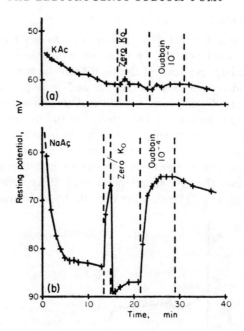
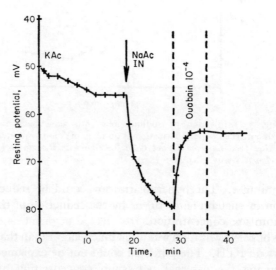

Fig. 16.—The hyperpolarization brought about by the injection of sodium ions into a snail neuron can be reduced by application of ouabain or reduction of K_o.

reducing potassium around the cell to zero level *reduces* the membrane potential instead of increasing it! (*See Figs.* 16, 48, 49.)

Thomas (1968) inserted four electrodes into a snail neuron. One electrode was made of sodium-sensitive glass so that he could measure the sodium concentration within the cell. The other electrodes were used to inject sodium ions at a known rate into the cell, to record the membrane potential and to current clamp the cell and so make it possible to measure the work done by the electrogenic sodium pump. Thomas showed that the current developed by the cell was proportional to the increase in internal sodium concentration (*Fig.* 17) above the normal level.

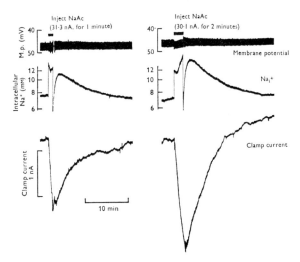

Fig. 17.—The hyperpolarization of a snail neuron following injection of sodium ions is proportional to the internal sodium concentration. Solid bars indicate the injection of NaAc. (*Reproduced from Thomas, R. C.* (1969), '*J. Physiol, Lond.*', **201**, 495–514.)

EFFECT OF TEMPERATURE ON NERVE-CELLS

The Nernst equation has a temperature coefficient (T) indicating that the membrane potential should change with temperature. It is simple to calculate what the change in membrane potential should be in nerve- or muscle-cells following a temperature change of 10° C. Kerkut and Ridge (1961, 1962) working on snail neurons and muscle-cells of crabs, insects, and frogs found that the experimentally determined membrane potential changes for 10° C. were much greater than those predicted by the Nernst equation.

Kerkut and Ridge (1961) came to the conclusion, 'The results reported in this paper do offer suggestive evidence that the resting potential of muscle fibres is a complex phenomenon linked with metabolic activity rather than a simple physical diffusion potential.'

These results were confirmed by Murray (1966) and by Carpenter (1967) for neurons and by Geduldig (1968a, b) in his work on muscle.

The study of the temperature effect was extended by Carptenter and Alving (1968) working on the large neurons of *Aplysia*. There was a much greater temperature sensitivity of these neurons than would be predicted from the Nernst equation. The equation predicted a change of 0·5 mV. per °C. whilst the experimentally observed value was up to 2 mV. per °C. The effect was reduced by ouabain or by adding lithium ions to the medium. The greater temperature sensitivity of the neurons was explained in terms of the greater sensitivity of the electrogenic sodium pump to temperature changes. The metabolic system was more sensitive than an equilibrium potential would be. In some *Aplysia* nerve-cells the electrogenic sodium pump can contribute 60 per cent of the membrane potential.

CRAYFISH STRETCH RECEPTOR

These large single neurons in the muscles of the crayfish provide a very useful preparation for micro-electrode and/or sensory physiology studies. Nakajima and Takahashi (1965, 1966a, b) showed that the crayfish stretch receptors gave a marked PTH of up to 11 mV. The recordings were by means of micro-electrodes. They were also able to inject sodium ions into the neurons and found that this would bring about a hyperpolarization. Estimations of the E_K showed that it had not changed during the PTH and they showed that dinitrophenol would reduce the PTH.

After a very thorough investigation of the changes in the membrane resistance during the PTH, they decided that the PTH could only be satisfactorily explained in terms of an electrogenic sodium pump.

There is now evidence that the electrogenic sodium pump can contribute to the membrane potential in various tissues, as shown in *Table IV*.

Table IV.—Contribution of the Electrogenic Sodium Pump to Membrane Potential

Species	Tissue	Author*
Mammal	Cat spinal moto-neuron	Coombs, Eccles, and Fatt (1955a, b); 10 mV.
		Ito and Oshima (1965); 5 mV.
	C fibres	Ritchie and Straub (1957); 4 mV.
	A fibres	Rang and Ritchie (1968); 35 mV.
Frog	Nerve-trunk	Connelly (1959); 8 mV.
	Spinal ganglion	Nishi and Soeda (1964); 4 mV.
Helix	Nerve-cells	Kerkut and Thomas (1965); 30 mV.
		Thomas (1968); 30 mV.
Aplysia	Nerve-cells	Carpenter and Alving (1968); 30 mV.
Sepia	Giant axon	Hodgkin and Keynes (1955); 1·8 mV.
		Shaw and Newby (1970); 14 mV.
Orconectes	Stretch receptor	Nakajima and Takahashi (1965, 1966a, b); 11 mV.
Other preparations		
Muscle	Frog muscle	Kernan (1962a, b); 12 mV.
		Adrian and Slayman (1964); 20 mV.
	Cat heart muscle	Tamai and Kagiyama (1968); 170 mV.
Neurospora	Not a sodium pump, but some other ion is pumped (H+)	Slayman (1965, 1970); 100 mV.

* The value in mV. indicates the possible contribution of the electrogenic sodium pump.

Electrogenic Sodium Pump in Muscle

The method of studying the electrogenic sodium pump in muscle is to isolate the muscle and store it for 12 hours at a low temperature in a potassium-free, high-sodium Ringer solution. The muscle is then removed and placed in normal Ringer at standard temperature.

Kernan (1962a, b) showed that muscles that had been loaded with sodium ions and then placed in a Ringer containing 10 mM K had a higher membrane potential than that calculated from the Nernst equation. Values up to 12 mV. greater than the calculated value were found.

If ouabain was added to the solution or if time was given for the sodium ions to be pumped out, the membrane potential came very close to the value predicted from the E_K.

Adrian and Slayman (1964) showed that the increased membrane potential found in sodium-loaded muscles could not be due to the uptake of potassium ions around the muscle. They replaced all the external potassium in the Ringer solution by rubidium. The sodium-loaded muscles showed a potential some 15 mV. greater than that predicted by the Nernst equation even though there were no external potassium ions and so no change in the E_K.

In heart muscle Tamai and Kagiyama (1968) showed that cooling the muscle to 4° C. for a period of time from 4 to 50 hours and then replacing the muscle in normal warmed Ringer solution led to a considerable increase in the membrane potential. Increases of up to 120 mV. were found. The control values for the membrane potential of the cat ventricular muscle was 87·3 mV. After cold treatment for 20 hours the potential on rewarming in normal Ringer was 267 mV. Ouabain, sodium azide, or dinitrophenol reduced this hyperpolarization.

Post-synaptic Electrogenic Effects

Several cases have been described where the transmitter chemical at the synapse stimulates the electrogenic sodium pump in the post-synaptic cells and so brings about a long-lasting hyperpolarization.

Frog Sympathetic Ganglion

Nishi and Koketsu (1967a, b) have analysed the synaptic responses in the frog sympathetic ganglion following orthodromic stimulation of the nerve-trunk. They recorded the ganglionic potentials by means of the sucrose-gap technique. Stimulation of the nerve-trunk led to an excitatory post-synaptic potential (EPSP) in the ganglion followed by a long-lasting hyperpolarization. This latter hyperpolarization was called the P potential.

The P potential appeared to be relatively independent of changes in the external potassium concentration. The P potential was abolished by ouabain whilst the EPSP was unaffected by ouabain. Dinitrophenol reduced the P potential. The P potential showed a greater sensitivity to temperature changes than that predicted from the Nernst equation. Nishi and Koketsu suggested that the P potential was due to the stimulation of the electrogenic sodium pump in these neurons and that this brought about the long-lasting hyperpolarization, the P potential.

DOPAMINE ON *Aplysia* NEURONS

In the abdominal ganglia of *Aplysia* are a series of large cells which are easily identified. One particular cell, the Br cell, showed a particular pattern of activity: a series of action potentials followed by a hyperpolarization and an inhibition that can last for several seconds. This type of activity is called 'inhibition of long duration' and cells that show it are called CILDA cells (*Fig.* 18). Injection of dopamine into the Br cells brings about a marked hyperpolarization (Ascher, 1968). Part of this hyperpolarization is reduced by ouabain. The reversal potential of this cell is higher than that of the dopamine response in adjacent cells. It is possible that part of the lasting hyperpolarization is due to stimulation of the electrogenic sodium pump in the neuron by dopamine.

DOPAMINE ON *Helix* NEURONS

Kerkut, Brown, and Walker (1969a, b) found a specific cell in the ganglion of *Helix* that was hyperpolarized by dopamine. The hyperpolarization could last for several seconds. Addition of ouabain at concentrations of 10^{-5} g. per ml. abolished the response to dopamine but rapid washing would restore the response. This hyperpolarization is thought to be due to dopamine stimulating the electrogenic sodium pump in the neuron.

ACETYLCHOLINE ON *Helix* NEURONS

Another identified neuron in the parietal ganglion of *Helix* showed a marked hyperpolarization to acetylcholine. The effect was inhibited by ouabain (Kerkut and others, 1969a, b). The cell's response was studied in more detail and compared with an adjacent cell that gave a classic hyperpolarization to acetylcholine but where this was due to an increase in the membrane permeability to chloride ions.

If a KCl-filled micro-electrode was inserted in the Type I cell, the ACh response gradually reversed as the chloride concentration of the cell increased. In the Type II cell injection of chloride ions had no effect on the ACh response. It remained a steady hyperpolarization (*Fig.* 19). Similarly the addition of ouabain had no effect on the ACh response of the Type II cell (*Fig.* 20). If the external sodium concentration around the ganglia was reduced, then the ACh response of the Type II cell was lost. It was restored by addition of 10 mM external sodium. Tetrodotoxin will also

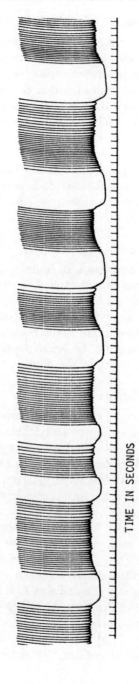

Fig. 18.—CILDA activity of *Aplysia* neurons. The neurons show an inhibition of longer duration than an IPSP.

THE NATURE OF SCIENTIFIC PROGRESS

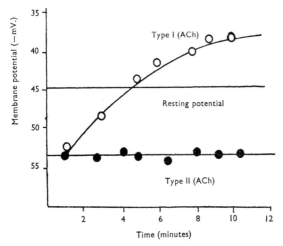

Fig. 19.—The effect of chloride injection on a Type I snail neuron (where the ACh effect is to increase the chloride permeability) and the Type II neuron. The ACh response of the Type I neuron is reversed by chloride injection, the Type II neuron is not affected. (*Figs.* 19, 20 *reproduced from Kerkut, G. A., Brown, L. C., and Walker, R. J.*, 1969b.)

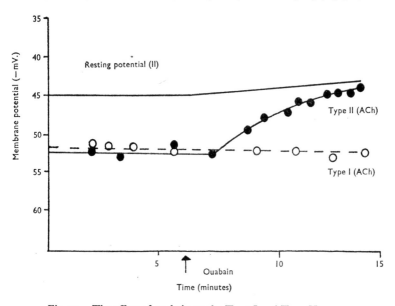

Fig. 20.—The effect of ouabain on the Type I and Type II neurons. The Type I neuron's response to ACh is not affected by ouabain. The Type II neuron's response to ACh is reduced, as is also its resting potential.

reduce the Type II response. It is suggested that the action of acetylcholine on the Type II cell is to increase the permeability of the membrane to sodium ions so that sodium slowly enters the cell. This stimulates the electrogenic sodium pump and the cell hyperpolarizes. If the cell is maximally stimulated by injecting sodium ions into the cell through a micro-electrode, then ACh has no visible effect on the membrane potential.

ACETYLCHOLINE ON *Aplysia* NEURONS

Pinsker and Kandel (1969) found an identified neuron (L10) in the abdominal ganglion of *Aplysia* which, when stimulated, set up post-synaptic potentials in a series of follower cells. In some of these follower cells, such as L2, there were two types of inhibitory potentials. There was an immediate quick inhibitory postsynaptic potential (IPSP) and a later longer IPSP.

There were many differences between these two types of IPSP, and Pinsker and Kandel suggested that the quick IPSP was the normal classic type of IPSP where there was an increase in the chloride conductivity of the membrane and that the later IPSP might be due to acetylcholine stimulating the electrogenic sodium pump. The differences between the two types of potential in the same cell are shown in *Table V*.

Table V.—DIFFERENCES BETWEEN TWO TYPES OF POTENTIAL IN THE SAME CELL

PROPERTY	EARLY IPSP	LATE IPSP
Cholinergic	Yes	Yes
Duration	Approx. 100 msec.	800+msec.
Curare	Blocked	No effect
Atropine	Blocked	No effect
Ouabain	No effect	Blocked
Hyperpolarize	Reversed	No effect
Reduce K_o	No effect	Blocked
Conductance	Change	No change
Reduce Cl_o	Reversed	No change
Temperature	Normal	Very sensitive

Kehoe and Ascher (1970) examined the results obtained by Pinsker and Kandel and concluded that the late IPSP is not due to an electrogenic sodium pump potential. Kehoe (1967, 1969)

THE NATURE OF SCIENTIFIC PROGRESS

had previously shown early and late IPSP's in *Aplysia* neuron-where the early phase was due to an increase in chloride permeability and the late IPSP due to a change in potassium permes ability. Kehoe and Ascher thought that Pinsker and Kandel's results could be similarly explained and suggested that ouabain, by stopping the Na–K pump, probably increased the external K concentration. This affected the E_K and reversal potential and at certain levels of membrane potential could be seen as an apparent loss of the IPSP. Kehoe and Ascher present a well-reasoned case for the early IPSP and late IPSP to be considered purely in terms of changes in membrane permeability and not in terms of an electrogenic sodium pump. (*See* p. 144 and *Fig.* 97.)

THE ROLE OF THE ELECTROGENIC SODIUM PUMP IN ANOXIA

Nerve-cells are very sensitive to anoxia and reduced Po_2.

Work on invertebrate neurons by Chalazonitis and his colleagues (Chalazonitis, 1961) has shown that the membrane potential of some neurons was very sensitive to changes in Po_2. The neurons of *Aplysia* have an internal haemoglobin-like pigment that indicates the Po_2 within the neuron. By simultaneously recording the dissociation of the pigment and the membrane potential he was able to show the changes of the membrane potential and Po_2 within the nerve-cell. If the Po_2 was reduced there was an immediate reduction in the membrane potential, whilst if the Po_2 was increased there was an immediate increase in the membrane potential (*Fig.* 21).

Kerkut and York (1969) showed that some of the membrane sensitivity to change in Po_2 could be explained in terms of the activity of the electrogenic sodium pump. They found that if they increased the Po_2 there was a hyperpolarization of the snail neuron and that it was prevented by ouabain or by scillaren (*Figs.* 22, 23).

Neurons were subjected to a standard change in Po_2 from 250 to 40 mm. Hg. One group of 58 cells had potassium injected inside them and they showed a membrane potential change of 4·6 ± 0·31 mV. to the change in Po_2. Another group of 57 cells had sodium injected inside them and they showed a membrane potential change of 8·6 ± 0·6 mV. to the change in Po_2.

The sodium-injected cells (whose electrogenic sodium pump had been stimulated) were much more sensitive to reduction in

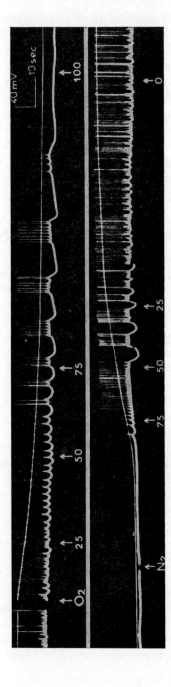

Fig. 21.—The effect of a reduction in Po_2 on the membrane potential of *Aplysia* neurons. The internal Po_2 is recorded by means of the absorption of the intracellular haemoglobin in the neurons. Reduction in Po_2 depolarizes the membrane. The numbers 25, 50, etc., on the smooth line indicate the internal oxygen concentration of the nerve-cell. (*Reproduced from Chalazonitis, N., Gola M., and Arvanitaki, A.* (1965), '*C. r. Séanc. Soc. Biol.*', **159**, 2451–2455.)

Po₂ than were potassium-injected cells. Similarly the sodium-injected cells showed the greatest potential change, up to 22·5 mV.,

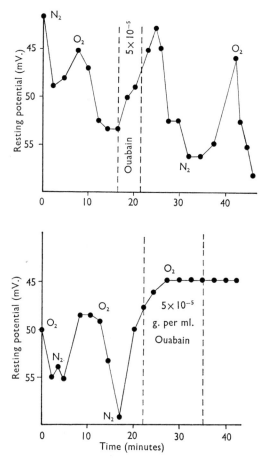

Fig. 22.—The effect of ouabain on the membrane sensitivity to Po₂. Ouabain reduces the sensitivity of the neuron to changes in Po₂, indicating that much of the sensitivity could be due to the electrogenic sodium pump being affected by Po₂. (*Figs.* 22, 23 *reproduced from Kerkut, G. A., and York, B.* (1969), '*Comp. Biochem. Physiol.*', **28**, 1125-1134.)

for the change in Po₂ whilst the most that the potassium-injected cells showed was 7·5 mV.

It seems likely that one of the first effects of anoxia on nerve-cells is to slow up the electrogenic sodium pump and thereby depolarize the neurons. This will then affect the synaptic and action potentials in those neurons.

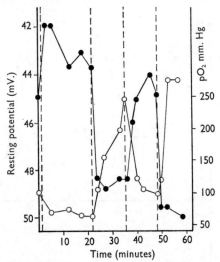

Fig. 23.—The effect of a reduction in Po_2 on the membrane potential and the activity of neurons in *Helix*. Reduction in Po_2 reduces the membrane potential. Increase in Po_2 increases the membrane potential.

CAT CORTICAL NEURONS

Krnjevic and Schwartz (1967a) found that certain neurons in the cat cerebral cortex were depolarized by acetylcholine. There was no conductance change during the application of acetylcholine. The response was inhibited by anoxia or by dinitrophenol. They suggested that the effect of acetylcholine could be to stimulate an electrogenic pump (electrogenic chloride pump?) which then depolarized the cell.

OTHER ROLES FOR THE ELECTROGENIC SODIUM PUMP

Nicholls and Baylor (1968) suggested that the glial cells in the leech could affect the level of potassium present around a neuron and so affect the electrogenic sodium pump of the neurons. The pump slows if the external potassium level is reduced (see p. 151).

Nakajima and Onodera (1969) suggest that the electrogenic sodium pump could be responsible for the adaptation of the stretch receptor of the crayfish to constant stimulus.

Fingerman (1969) has evidence that the chromatophores of *Leander* are inhibited by ouabain and that a sodium–potassium

THE NATURE OF SCIENTIFIC PROGRESS 55

pump can affect the membrane potential of the cell and so affect the dispersion of the chromatophore.

SUMMARY OF THE SIGNIFICANCE OF THE ELECTROGENIC SODIUM PUMP

1. The sodium–potassium pump in nerve- and muscle-cells can be electrogenic.
2. The precise value of the potential developed depends upon:—
    a. The rate of action of the pump.
    b. The increase in internal sodium concentration.
    c. The membrane resistance.
    d. The membrane potential.
    e. The external potassium concentration.
    f. The availability of ATP.
3. In some identified neurons of *Helix* and *Aplysia*, where the membrane potential is about 50 mV., up to 20 mV. of this are attributable to the electrogenic sodium pump.
4. The electrogenic sodium pump is inhibited by ouabain, dinitrophenol, lithium, or reduction in K_o.
5. The electrogenic sodium pump potential is three to four times more sensitive to temperature change than is the membrane potential dominated by E_K.
6. The electrogenic sodium pump is stopped by low Po_2 or anoxia.
7. Certain synapses show post-synaptic stimulation of the electrogenic sodium pump by acetylcholine or dopamine. This results in a long-lasting hyperpolarization and inhibition.
8. Regulation of K_o by the glial cells of the leech could affect the electrogenic sodium pump of adjacent neurons and so control their membrane potential.
9. The electrogenic sodium pump could contribute to the sensory adaptation of crustacean stretch receptors.
10. It is possible that there are other electrogenic pumps. An electrogenic chloride pump has been suggested for cortical neurons. It is also possible that the linkage between the Na : K could vary and this would affect the membrane potential.
11. The electrogenic sodium pump provides the neuron with a mechanism by which metabolism can control the critical level of the membrane potential of the cell.

The preceding pages have given a general account of the problems involved in understanding the unequal distribution of ions across membranes, the electrical potentials across membranes, and the relationship of the metabolism of the cell to these.

Part II will present the evidence in more detail concerning the relationship between the active movement of sodium ions across membranes and the development of a potential across the membrane, i.e., the electrogenic sodium pump.

PART II

THE ELECTROGENIC SODIUM PUMP IN NERVE AXONS

Sodium Injection into Squid Giant Axons

It was suggested by Hodgkin and Keynes (1955) that Na^+ transport from squid axons could be electrogenic. In their earliest work on active Na transport they loaded squid axons with labelled Na^+ and then measured the rate of Na^+ extrusion. They stimulated squid axons in ^{24}Na Ringer to load them and drew them into short lengths of capillary tubing after washing off the extracellular ions. Radioactivity was collected from the length of the fibre, but not from the ends, by withdrawing fluid from a side-arm in the capillary with a motor-driven syringe (*Fig.* 24). Using this

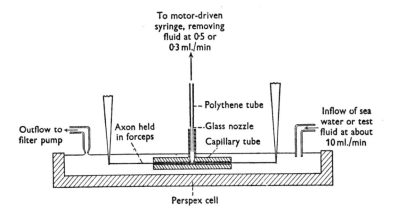

Fig. 24.—Apparatus used for measuring radioactive Na efflux from squid and *Sepia* axons. (*Figs.* 24, 25 *reproduced from* Hodgkin, A. L., and Keynes, R. D. (1955), '*J. Physiol., Lond.*', **128**, 28–60.)

method, Hodgkin and Keynes showed that the removal of K from the bathing sea water reduced the Na efflux to about one-third of the previous value, with a lag of only 2–3 minutes. In contrast, inhibition of Na^+ efflux by metabolic inhibitors such as 2,4-dinitrophenol shows a longer delay (*Fig.* 25). Since one-third of the Na efflux was not coupled to K efflux, Hodgkin and Keynes suggested that this Na could be ejected as ions and generate a potential. The measured uncoupled Na movement would generate a current of 1·8 mA., but as the resting membrane resistance is only 1000 Ω per sq. cm. the direct contribution to the membrane potential would be only 1·8 mV., an insignificant amount. The lack of any marked change in membrane potential on applying 2,4-dinitrophenol to the squid axon is therefore not conclusive evidence that the pump is neutral.

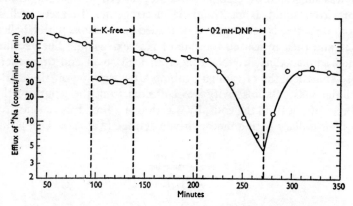

Fig. 25.—Effect of K-free sea water and 0·2 mM DNP on sodium efflux from a *Sepia* axon. K-free Ringer immediately reduces Na efflux to one-third of its initial level. 0·2 mM DNP takes longer to act.

A year later Hodgkin and Keynes (1956) used a microsyringe to inject a number of electrolytes, including NaCl, into squid giant axons. This syringe ejected a volume of fluid equal to its own external volume when it was withdrawn from the axon. This arrangement prevented any damage to the axon by volume changes.

Injection of 1·0 M NaCl caused a slight increase in membrane potential. In 4 of 5 experiments the membrane potential rose by 1 to 4·5 mV. NaCl injection also stimulated the efflux of Na. If some of this Na efflux was uncoupled to K influx the increase in membrane potential on Na injection could be due to stimulation of electrogenic Na efflux by the increase in intracellular Na.

Grundfest, Kao, and Altamirano (1955) had already suggested that an electrogenic Na pump contributed to the membrane potential of squid giant axons. Results obtained from micro-injection of ions into these axons led Grundfest to conclude that the resting potential could not be explained in terms of a Gibbs-Donnan equilibrium. On finding that the resting potential was insensitive to changes of K or Cl, he suggested instead that the resting potential was generated by a pump mechanism which ejected Na.

Post-tetanic Hyperpolarization in Mammalian Non-medullated Nerve-fibres

Discovery

Some comprehensive evidence for the existence of an electrogenic Na pump was provided by Ritchie and Straub (1957), although at that time this evidence was not interpreted as being due to such a pump. They worked on rabbit non-medullated

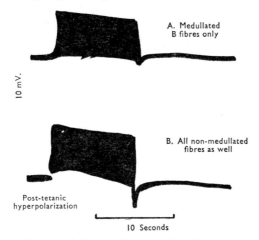

Fig. 26.—Post-tetanic hyperpolarization in rabbit nerve-trunk, following stimulation of small diameter C fibres. (*Figs.* 26–29 *reproduced from Ritchie, J. M., and Straub, R. W.* (1957), '*J. Physiol., Lond.*', **136**, 80–97.)

nerve-fibres of small diameter in the rabbit's cervical sympathetic trunk. This trunk contains both medullated B fibres of large diameter and non-medullated C fibres of small diameter. Repetitive stimulation of the nerve-trunk led to a post-tetanic hyperpolarization which was confined to the C fibres (*Fig.* 26). After a 10-second period of stimulation at a low intensity, just sufficient

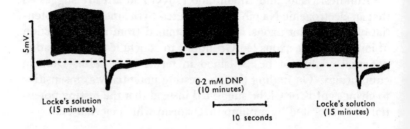

Fig. 27.—Post-tetanic hyperpolarization in nerve is greatly reduced by 0·2 mM DNP. The effect is reversed by washing.

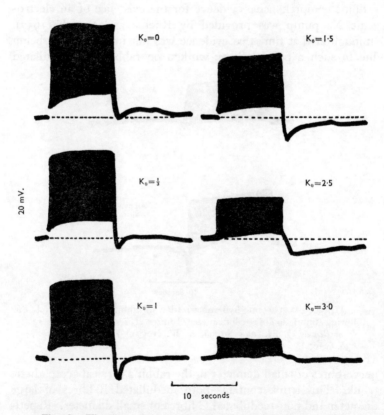

Fig. 28.—The size and time course of post-tetanic hyperpolarization in nerve are very dependent on the external K concentration. They are reduced in amplitude by both low (0–0·3 mM) and high (3 mM) K. High K also prolongs the time course of the hyperpolarization.

to stimulate the larger diameter B fibres, there was little or no hyperpolarization (*Fig.* 26 A). However, following an increase in intensity which stimulated the C fibres as well as the B fibres, there was a hyperpolarization in the order of 4 mV. (*Fig.* 26 B).

2,4-DINITROPHENOL, OUABAIN, AND POTASSIUM CONCENTRATION

After soaking a desheathed sympathetic trunk in 0·2 M 2,4-dinitrophenol for 15 minutes the post-tetanic hyperpolarization shown by the non-medullated fibres was very much reduced, and the membrane potential after tetanus was depolarized by about 0·6 mV., as shown in *Fig.* 27. Ten minutes after restoring the preparation to Locke's solution the post-tetanic hyperpolarization had nearly returned to the initial size of 3 mV. (*Fig.* 27). Ouabain (15 μM) also removed the hyperpolarization.

Fig. 28 shows the effect of varying the external K concentration on the potential recorded during and after a 10-second period of repetitive stimulation. Both the size and time course of the response were influenced by the external K concentration. The rate of onset and of delay were slowed by high external K (3·0 mM), but in very low concentrations of external K (0–0·3 mM) the amplitude of the post-tetanic hyperpolarization was greatly reduced.

LITHIUM IONS

Ritchie and Straub (1957) concluded from these results that post-tetanic hyperpolarization was associated with some energy-requiring process. Na extrusion seemed the most likely candidate, since Na enters the nerve during the action potential. This idea was supported by the fact that, after soaking the nerve-bundle for 4 minutes in a solution where all the NaCl has been replaced by LiCl, the post-tetanic hyperpolarization after a 10-second period of stimulation was completely removed (*Fig.* 29 B). The effect was reversed on returning to Na-containing Locke's solution, as shown in *Fig.* 29 C. Since Li inhibits the Na–K pump (Keynes and Swan, 1959) it appears that the hyperpolarization is associated with the activity of such a pump.

Ritchie and Straub suggested that tetanus caused Na-loading of the fibres which led to a stimulation of an electroneutral Na–K linked pump. If the K was taken up into the muscle-cells at a faster rate than it could diffuse into the extracellular spaces from the bulk of the bathing solution, the equilibrium potential for K, and thus the membrane potential, would increase. The hyperpolarization would be due to the rise in internal K in the nerve-trunk and the fall in external K immediately around the axon. Following the Nernst equation for E_K, the membrane potential would increase. They did not think that the hyperpolarization was caused by stimulation of an electrogenic Na pump in response to the raised internal Na, as there was little evidence for such a pump in other tissues.

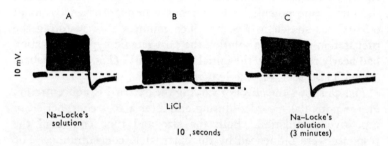

Fig. 29.—Effect of Li on post-tetanic hyperpolarization in rabbit nerve. Replacement of external Na by Li removes post-tetanic hyperpolarization. The effect is reversed by replacing Na.

POST-TETANIC HYPERPOLARIZATION OF FROG NERVE-FIBRES

A careful examination of the properties of the post-tetanic hyperpolarization in nerve-fibres led both Connelly (1959) and Straub (1961) to propose that this process was due to the stimulation of an electrogenic Na pump.

Connelly (1959) found that post-tetanic hyperpolarization in A fibres in the frog had properties similar to those found by Ritchie and Straub (1957) in mammalian non-medullated fibres. He tetanized the frog nerve, interrupting the tetanus for 5 seconds once every minute. At the end of the tetanus the post-tetanic hyperpolarization lasted for more than 1 hour, as shown in *Fig.* 30. The positive deflexions shown in the figure are hyperpolarizations at each period of interruption of tetanus.

The rise in membrane potential after tetanus was very sensitive to changes in external K. The hyperpolarization following a stimulation of 25 per minute in 4 mM K increased in high K concentrations (8·5 mM) and was very much decreased and prolonged in K-free Ringer (*Fig.* 31). Replacement of Na by Li eliminated the hyperpolarization.

Connelly concluded that post-tetanic hyperpolarization in nerve was due to uncoupled Na extrusion rather than to a neutral pump, since the following experiment could not be explained by the argument that the hyperpolarization was due to a raised internal K and a decreased external K concentration. He stimulated the frog A fibres at 10 per second in 5 mM K Ringer and recorded the membrane potential. During the tetanus the average membrane potential should have been less or the same as that before stimulation if a neutral pump was involved, since there would be a net

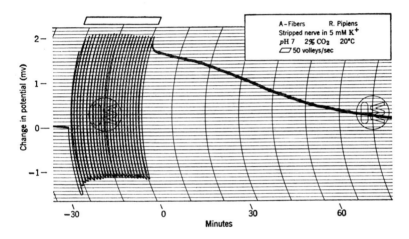

Fig. 30.—Post-tetanic hyperpolarization in frog nerve. Tetanus in interrupted for 5 seconds every minute. After tetanus the hyperpolarization lasted for more than 1 hour. (*Figs.* 30–32 *reproduced from* Connelly, C. M. (1959), '*Rev. mod. Phys.*', **31**, 475–484.)

loss of K from the fibres. However, the membrane potential became hyperpolarized 4 minutes after the beginning of tetanus and remained so throughout the period of stimulation as shown in *Fig.* 32. The same results were obtained in crab nerve during a 43-minute period of tetanus at 1 volley per second.

Connelly put forward a model to describe the shape of the recovery curves of the membrane potential after tetanus for various periods of time. His analysis supported the hypothesis that the hyperpolarization was due to a saturable, ion-extruding mechanism, operating at the nodes of the frog nerve (i.e., an electrogenic Na pump).

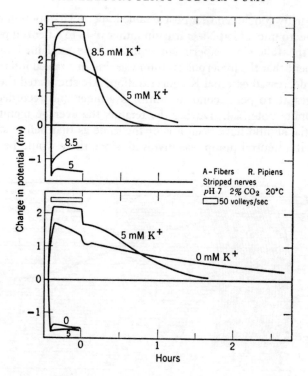

Fig. 31.—Effect of external K on post-tetanic hyperpolarization in frog nerve. The recovery is rapid in high K (8·5 mM) and is very slow and prolonged in K-free solutions.

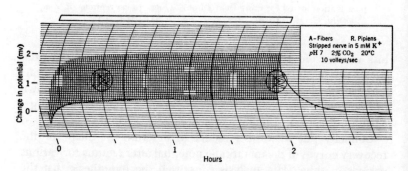

Fig. 32.—Changes in the average membrane potential of frog nerve during and after tetanus. Tetanus was interrupted for 5 seconds every minute. The membrane potential increased even *during* tetanus.

Straub (1961) studied the mechanism of post-tetanic hyperpolarization production following activity in small diameter myelinated nerve-fibres from frog using a sucrose-gap technique. He showed that the amplitude of the post-tetanic hyperpolarization was directly proportional to the membrane resistance. If the post-tetanic hyperpolarization was dependent on an increase in the value of E_K by a neutral pump, then an *inverse* relationship between membrane resistance and amplitude of post-tetanic hyperpolarization would be expected. However, if post-tetanic hyperpolarization was due to an electrogenic Na pump the potential generated would be directly proportional to the membrane resistance.

The membrane potential during post-tetanic hyperpolarization could exceed the membrane potential in K-free solution, sometimes by as much as 9 mV. An estimation of the potential in K-free Ringer, allowing for a short-circuiting factor, showed that the post-tetanic hyperpolarization was too large to be explained in terms of a neutral pump. Straub also estimated that the increase in internal Na during tetanus was too small to produce the observed hyperpolarization by coupled Na–K transport. A similar calculation suggested that the hyperpolarization could be explained if only half the active Na efflux were electrogenic.

Later studies on post-tetanic hyperpolarization in mammalian non-medullated nerve-fibres (Greengard and Straub, 1962) and frog tibial nerve (Bohm and Straub, 1961) have shown that anoxia and a wide variety of metabolic inhibitors abolish post-tetanic hyperpolarization.

REMOVAL OF SHORT-CIRCUITING BY CHLORIDE

More recent evidence produced by Rang and Ritchie (1968) has confirmed that post-tetanic hyperpolarization is due to the stimulation of an electrogenic Na pump. A problem in the experiments described so far is that only slight potential changes were involved. The reason for this turned out to be that Cl was leaving the cells passively with the active Na efflux and short-circuiting the pump, so that only a small rise in potential was recorded.

Working with mammalian non-medullated nerve-fibres, Rang and Ritchie (1968) obtained an initial fast and a later slow component of hyperpolarization after repetitive stimulation of the nerve-trunk. The early component was probably due to an increase in K permeability. They made a study of the later component. In normal Locke solution the nerve-trunk was hyperpolarized by 2–4 mV. after stimulation for 5 seconds at 30 per second.

If any uncoupled outward movement of Na after tetanus was accompanied by a passive efflux of Cl, the potential generated by the Na current would be reduced. In the absence of internal Cl the Na should leave the cell unaccompanied by an anion and so generate a higher potential. The internal Cl was depleted by soaking the preparation in a Ringer solution in which the external Cl had been replaced by a large impermeable anion such as isethionate. When the Cl of the Locke solution was replaced by isethionate and the nerve was again stimulated for 5 seconds at 30 per second, the hyperpolarization after tetanus was increased from 2 to 4 mV. up to about 20 mV., the maximum hyperpolarization obtained being 35 mV. This effect of isethionate is shown in

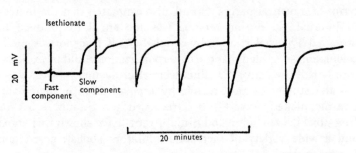

Fig. 33.—Effect of replacing external Cl by isethionate on posttetanic hyperpolarization in rabbit nerve. The post-tetanic hyperpolarization increased from 1–2 mV. in Cl solution to 20 mV. in isethionate solution. (*Figs.* 33–37 *reproduced from Rang, H. P., and Ritchie, J. M.* (1968), '*J. Physiol., Lond.*', **196**, 183–221.)

Fig. 33. The fast and slow components in unmodified Locke solution are indicated by arrows. On admission of isethionate at the arrow the size of the hyperpolarization following identical stimuli gradually increased and reached a maximum after 10–15 minutes, presumably when the internal Cl was depleted. The large size of the hyperpolarization makes it almost certain that an electrogenic Na pump was responsible.

Fig. 34 illustrates the effect of ouabain on post-tetanic hyperpolarization in isethionate–Locke solution. When 1 mM ouabain was applied to the nerve immediately after tetanus, the post-tetanic hyperpolarization was reduced in length (*Fig.* 34 A, B) and later hyperpolarizations were completely abolished (C).

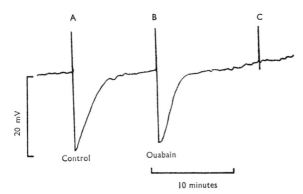

Fig. 34.—Effect of ouabain on post-tetanic hyperpolarization in rabbit nerve in the presence of isethionate. The post-tetanic hyperpolarization was completely abolished by ouabain.

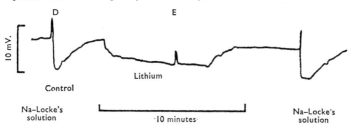

Fig. 35.—Effect of Li on post-tetanic hyperpolarization in rabbit nerve in the presence of isethionate. Replacement of external Na by Li reversibly inhibited the post-tetanic hyperpolarization.

When Li_2SO_4 was substituted for Na_2SO_4 in a sulphate–Locke solution the post-tetanic hyperpolarization was reversibly inhibited. *Fig.* 35 D and F show the response to tetanus in Na-containing Ringer. In *Fig.* 35 E there is no sign of a hyperpolarization when the preparation is in Li Ringer.

The amplitude and duration of the hyperpolarization in the C fibres were dependent on the external K concentration. When the nerve was stimulated in K-free isethionate the initial hyperpolarization ended suddenly. When K_2SO_4 was readmitted after the hyperpolarization had almost disappeared, a further hyperpolarization was seen almost immediately. *Fig.* 36 shows the effect of readmitting K in concentrations of 1, 2, and 5 mM. On readmission of 5 mM K the membrane potential was temporarily increased by over 20 mV. (*Fig.* 36 A). When 1 mM ouabain was added after electrical stimulation, but before the addition of K,

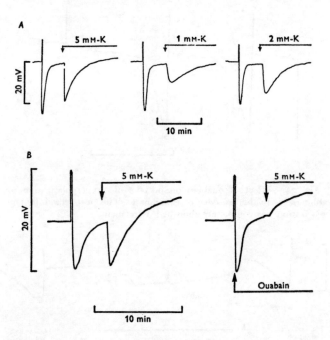

Fig. 36.—Activation of the recovery mechanism in rabbit nerve after tetanus by K and its inhibition by ouabain (1 mM). The posttetanic hyperpolarization is cut short in K-free Ringer. Admission of K 4 minutes later caused a large hyperpolarization which was abolished by ouabain.

the K-induced hyperpolarization was replaced by a depolarization. Such results would be hard to explain if the post-tetanic hyperpolarization was dependent on E_K. They are readily explained in terms of an electrogenic Na pump with a dependence on the presence of external K.

Rang and Ritchie also studied the ability of other cations, known to stimulate the Na pump in other tissues, to activate the recovery mechanism in the non-myelinated fibres after stimulation in K-free Locke solution. The activated responses shown in *Fig.* 37 are expressed as fractions of the initial hyperpolarizations in

K-free isethionate solution. Closed circles show the amplitude of responses to different concentrations of K. Open circles show

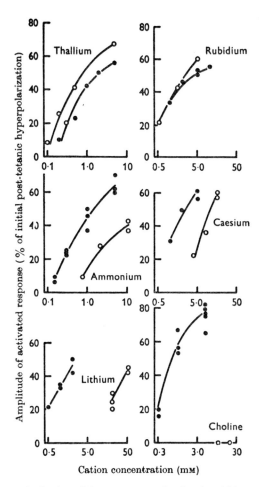

Fig. 37.—Activation of the recovery mechanism in rabbit nerve after tetanus by K and other cations. ●, Amplitude of responses to different K concentrations. ○, Corresponding responses to other cations. The initial post-tetanic hyperpolarization was in K-free Ringer.

corresponding responses to other cations. Thallium, rubidium, ammonium, caesium, and choline ions were all able to replace K in activating the recovery mechanism to some extent. Thallium was three times as effective as K, as can be seen from *Fig.* 37.

pH ON POST-TETANIC HYPERPOLARIZATION

Holmes (1962) also found that post-tetanic hyperpolarization in mammalian non-medullated nerve-fibres was very small at normal pH values. However, a low pH medium (pH 6·4) caused an irreversible increase in the post-tetanic hyperpolarization, without changing any of its properties. Holmes suggested that, although it was hard to explain how an *electroneutral* pump could be improved by low pH, the increase in hyperpolarization could be explained in terms of an electrogenic Na^+ pump if H^+ reduced the Na–K coupling. This explanation is supported by the fact that H^+ decreases membrane K^+ permeability in other tissues (Hagiwara, Gruener, Hayashi, Sakata, and Grinnel, 1968).

THE EFFECT OF METABOLIC INHIBITORS ON POST-TETANIC HYPERPOLARIZATION

A study has been made on the effect of metabolic substrates on the PTH in mammalian non-myelinated nerve-fibres (den Hertog,

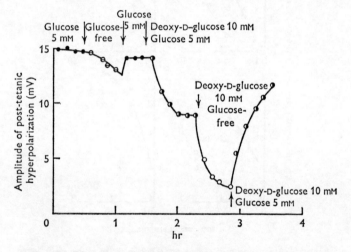

Fig. 38.—The effect of deoxy-D-glucose (DDG) on the amplitude of PTH. Glucose-free solution slightly reduced the PTH. DDG reduced the amplitude still more, the effect being reversible by addition of glucose. (*Figs.* 38, 39 *reproduced from den Hertog, A., Greengard, P., and Ritchie, J. M.* (1969), '*J. Physiol., Lond.*', 204, 511–522.)

Greengard, and Ritchie, 1969; den Hertog and Ritchie, 1969). They found that, although removal of glucose from the surrounding solution had no effect on the PTH, if an inhibitor of glucose, deoxy-D-glucose, was added to the solution, then the PTH was reduced or abolished (*Fig.* 38). The effect could be reversed by glucose, fructose, pyruvate, or acetate. The inhibitory effect of deoxy-D-glucose was enhanced by oxaloacetate and by malate.

Malonate reduced or abolished the PTH and this effect could be overcome by glucose and by pyruvate (*Fig.* 39).

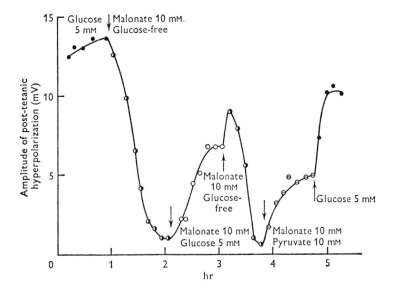

Fig. 39.—The effect of malonate on amplitude of PTH. Malonate reduced the amplitude of PTH. The effect of malonate was reversed by glucose or pyruvate.

The time course of the effect of metabolic inhibitors was similar to that obtained by cooling and differed from that obtained by the addition of ouabain. It seemed that ouabain was acting at a different site from that vulnerable to metabolic inhibitors.

PART III

THE ELECTROGENIC SODIUM PUMP IN NERVE-CELLS

The Electrogenic Sodium Pump in Motoneurons

SOME of the earliest evidence suggesting the existence of an electrogenic Na pump in nervous tissue was obtained by Coombs, Eccles, and Fatt (1955a, b), who used double-barrelled electrodes to inject ions into cat motoneurons. Injection of K into these cells had no effect, although it was not possible to alter the intracellular K concentration appreciably by this technique. In contrast, injection of 20-50 pico-equivalents Na nearly always diminished the membrane potential by about 10 mV. The reason for this depolarization appeared to be that the injected Na displaced the internal K and lowered the E_K, since the amplitude of the after potential (which is approximately equal to E_K) was considerably reduced after Na injection. The membrane potential returned to normal 8-9 minutes after Na injection. The membrane potential hyperpolarized by 10-15 mV. above the normal resting potential 10-20 minutes after Na injection. This rise in membrane potential was not accompanied by an appreciable resistance change. Coombs and others suggested that the hyperpolarization was due to the specific active extrusion of Na^+-carrying positive charges from the cell.

Ito and Oshima (1965) also found that Na injection into cat spinal motoneurons caused a large depolarization and they thought that this was due to a depletion of intracellular K. This suggestion was supported by the fact that the after-potential of these cells became a prolonged depolarization following Na injection. They found that the depolarization due to Na injection showed a marked membrane potential dependence. For example, at a membrane potential of −70 mV. K and Rb injection had about the same effect on the membrane potential as at − 80 mV. However, Na or Li injection resulted in a far greater depolarization at −70 mV. than at − 80 mV. It was suggested that Na injection stimulated an electrogenic Na pump which compensated for the depolarizing action of internal K depletion. The permeability of the membrane to K decreases with increasing membrane potential. Thus the contribution of the depolarizing effect of the reduced internal K is small compared with the hyperpolarizing effect of an electrogenic Na pump when the membrane potential is increased from 70 to 80 mV.

Kuno, Miyahara, and Weakly (1970) examined the post-tetanic hyperpolarization (PTH) in the dorsal spinocerebellar tract neurons of the cat. The amplitude and duration of the PTH were dependent on the number and frequency of the action potentials. There was no detectable conductance change during the PTH and there was no reversal potential for the PTH. The temperature coefficient (Q_{10}) was 2·4 within the range of 31–40° C. It was suggested that the PTH was produced by an electrogenic sodium pump in the nerve-cells.

THE ELECTROGENIC SODIUM PUMP IN SNAIL NEURONS
NaAc AND KAc INJECTION

The first workers to provide a simple demonstration of an electrogenic Na pump were Kerkut and Thomas (1965). They

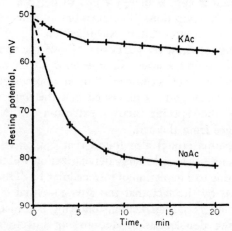

Fig. 40.—The effect of injecting either potassium acetate or sodium acetate on the resting potential. KAc has only a slight effect on the resting potential whilst NaAc has a marked effect, bringing the potential from 51 mV. to 82 mV. in 20 minutes. (*Figs.* 40–45 *reproduced from* Kerkut, G. A., *and* Thomas, R. C. (1965), '*Comp. Biochem. Physiol.*', **14**, 167–183.)

injected Na into snail neurons by means of low-resistance diffusion electrodes. *Fig.* 40 shows the effect of injecting either Na or K into snail neurons. The graph for KAc is the mean of results from 11 different experiments, while that for NaAc is the mean of 8 results. Injection of KAc caused an increase in average membrane potential from 51 to 58 mV. over 21 minutes. However, when NaAc was injected the membrane potential rose from an estimated value of 51 mV. to 80 mV. in 10 minutes.

The difference between the two responses was not due to an inherent difference between the nerve-cells studied, as can be seen from *Fig.* 41. First a KAc electrode and then a NaAc

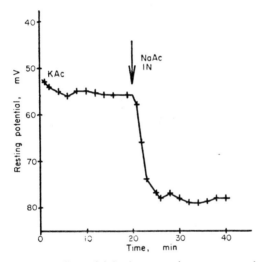

Fig. 41.—The effect of injecting potassium acetate and sodium acetate into the same cell. The KAc electrode was inserted at zero time and at 20 minutes a NaAc electrode was inserted. Notice the rise in the resting potential after insertion of the NaAc electrode.

electrode were inserted into the *same* cell. After insertion of the KAc electrode the membrane potential rose from 52 to 56 mV. over 20 minutes. After insertion of the NaAc electrode the membrane potential rose rapidly, reaching 78 mV. within 10 minutes.

To estimate the rate of diffusion of Na ions from the electrodes, NaCl instead of NaAc was injected into the IPSP cells. As the E_{Cl} of these cells has been shown to be related to the ACh reversal potential (Kerkut and Thomas, 1963, 1964) measurements of the rate at which the ACh potential altered would indicate the rate at which Cl, and thus Na, is diffusing from the micro-electrode. The value of the ACh potential was measured at intervals by applying high concentrations of ACh to the cell, since the membrane potential reached during application of high concentrations of ACh has been shown to be close to the ACh reversal potential. Average rates of diffusion of 4·5 mmole Na^+ per cell per minute were obtained from 9 experiments.

Effect of Lithium Injection

Injection of LiCl brought about only a temporary hyperpolarization in contrast to the larger sustained hyperpolarization caused by NaCl injection (*Fig.* 42). During NaCl injection the

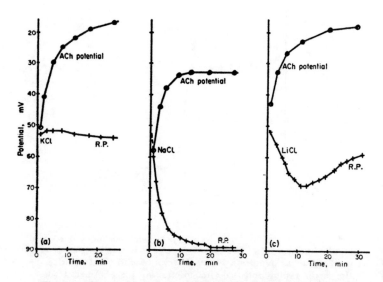

Fig. 42.—Comparison of the effect of injecting KCl, NaCl, and LiCl into nerve-cells. The ACh potential is related to the concentration of chloride ions in the cell and indicates the rate at which the Cl diffuses from the micro-electrode. Note that, though the salts are diffusing from all three electrodes, the K electrode has little effect on the membrane potential; the Na electrode brings about a marked hyperpolarization which is maintained; the Li electrode brings about a temporary hyperpolarization.

membrane potential rose rapidly from 50 to 86 mV. over the first 10 minutes and remained at a stable level of 89 mV. In contrast, injection of LiCl caused a rise in membrane potential from 48 mV.

to a peak of 68 mV. in the first 10 minutes, but then a slow decline of 59 mV. after 30 minutes.

OUABAIN AND K-FREE RINGER

Ouabain abolished the rise in membrane potential caused by Na injection. *Fig.* 43 shows that the rise in membrane potential from 56 to 79 mV. caused by Na injection was reduced by addition of 10^{-4} g. per ml. ouabain, the membrane potential falling to 64 mV. in 5 minutes.

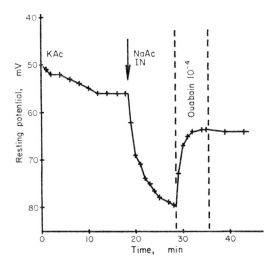

Fig. 43.—The effect of ouabain on the Na-induced hyperpolarization. The potential was recorded through a KAc electrode. When a NaAc electrode was inserted there was a hyperpolarization which was reduced on addition of ouabain.

The removal of external K also caused a fall in potential during Na injection but not during K injection, as shown in *Fig.* 44. While the reduction of external K to zero had no significant effect on the membrane potential when K was injected into the cell, it caused a 17-mV. drop in membrane potential over 1·5 minutes when Na was injected. The effect was readily reversible.

The rise in membrane potential was also removed by the SH-group inhibitor, parachloromercuribenzoate (PCMB). As shown in *Fig.* 45, $2·5 \times 10^{-5}$ g. per ml. PCMB caused a reduction of the Na-induced hyperpolarization of 14 mV.

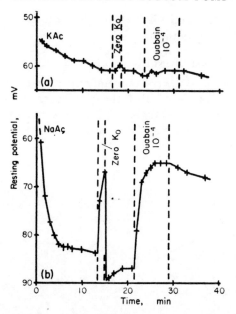

Fig. 44.—The effect of removing external K and of ouabain on the resting potential. Removing external K or adding ouabain had little effect on the cell when KAc diffused in (a) but had a marked effect on the NaAc-induced hyperpolarization (b). Note that reduction of external K, like ouabain, causes a fall in the hyperpolarization.

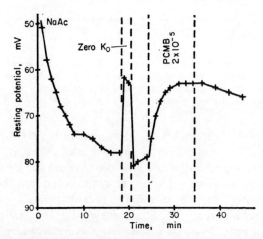

Fig. 45.—The effect of removing the external K and of PCMB on the Na-induced hyperpolarization. Removing external K or addition of PCMB (parachloromercuribenzoate) led to a fall in the Na-induced hyperpolarization.

The effect of Na injection on the membrane potential of snail neurons appears to be due to the activity of a Na pump since it is readily removed by ouabain and the inhibitor PCMB. The hyperpolarization cannot be due to a depletion of K just exterior to the cell by a neutral pump raising the E_K, as if this was the case K-free Ringer should have little effect or raise the Na-induced hyperpolarization. However, it severely reduced the rise in membrane potential. The evidence described strongly supports the presence of an electrogenic Na pump in snail neurons which is stimulated by a raised internal Na^+ concentration and is dependent on the presence of external K.

ELECTROGENIC SODIUM PUMP AND THE CONSTANT-FIELD EQUATION

Moreton (1969) extended the findings of Kerkut and Thomas (1965) on the electrogenic Na^+ pump in snail neurons. He allowed for the contribution of the electrogenic Na pump to the

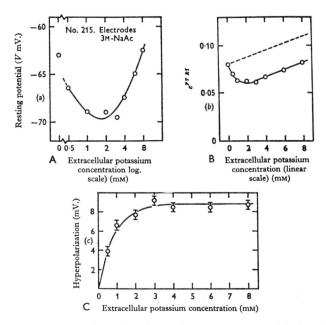

Fig. 46.—Potassium dependence of the resting potential of snail neuron during Na injection, plotted on a logarithmic (A) and linear (B) scale. Broken line (B) shows the theoretical behaviour, after subtracting the hyperpolarization due to the Na pump, the behaviour of which is shown in C. (*Figs.* 46, 47 *reproduced from Moreton, R. B.* (1969), '*J. exp. Biol.*', **51**, 181–201.)

membrane potential in a modification of a previous equation, derived from the constant-field theory (Moreton, 1968). The modified equation is as follows:—

$$E = \frac{FV}{e_{RT}} - \frac{(K_o)}{(K_i)} + \frac{P_{Na}(Na_o)}{P_K(K_i)} + \frac{RTM}{FVP_K(K_i)},$$

where M is the rate of net efflux of monovalent cations from the cell.

The behaviour of the membrane potential in response to changes in K_o during Na injection is shown in *Fig.* 46. Above 4 mM K the membrane potential showed a linear relationship to

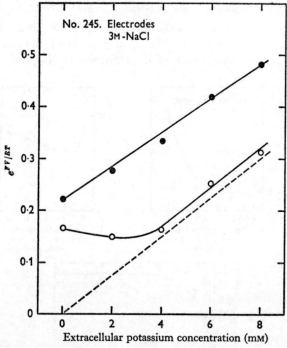

Fig. 47.—Potassium dependence of the resting potential of neuron during Na injection, immediately before (○) and 20 minutes after (●) exposure to ouabain. Broken line indicates E.

the external K. Below 4 mM the membrane potential was often less negative than the value required for a linear relationship. Moreton suggests that, as described in the above equation, the membrane potential during Na injection is affected by the rate of action of the electrogenic Na pump. He calculated the contribution made by the pump to the membrane potential as a function of the external K. *Fig.* 46 C shows the result of this calculation. Below 4 mM K the pumping rate falls considerably and decreases the contribution to the membrane potential, but above this concentration the rate is independent of external K. The deviation in linearity below 4 mM was abolished by 10^{-4} M ouabain (*Fig.* 47).

DEPENDENCE ON EXTERNAL POTASSIUM

Marmor and Gorman (1970) found that when the external potassium concentration around the neurons of *Anisodoris nobilis* was reduced, the membrane potential decreased at very low K_o

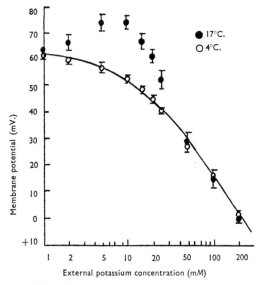

Fig. 48.—Effect of temperature on the potassium dependence of the resting potential of a snail neuron. The contribution of the electrogenic Na pump to the membrane potential can be seen between about 2 and 30 mM K at 17° C. (*Figs.* 48, 49 *reproduced from Marmor, M., and Gorman, A. L. F.* (1970), '*Science, N.Y.*', **167**, 65–67.)

(*Fig.* 48). These neurons also had their membrane potential reduced when ouabain was applied (*Fig.* 49). It would appear that the resting neurons had a component of about 10 mV. due to the electrogenic sodium pump. This component could be abolished either by addition of ouabain or by the removal of the external potassium concentration. In both cases there was a fall of 10 mV. in the resting potential.

Gorman and Marmor (1970a, b) give a more detailed account of their studies on the membrane potential of the neurons of the gastro-oesophageal ganglia of *Archidoris nobilis*.

The ganglion is small (300–500 μ in diameter) and nearly 50 per cent of its content is occupied by one neuron cell-body, the G cell, which is 200–400 μ in diameter. At 11–13° C. the cells

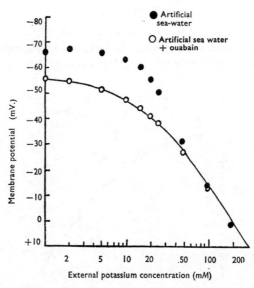

Fig. 49.—Effect of ouabain on the K dependence of the resting potential of snail neuron. Ouabain abolishes the contribution of the electrogenic Na pump to the curve.

had a resting potential of 50–60 mV. and action potentials of 100–120 mV. The cell does not show any pacemaker activity or synaptic potentials.

The response of the membrane potential to rapid changes in external K concentration was prompt, stable, and reversible up to 200 mM K. It was independent of the external Cl concentration.

Warming the cell produced a prompt hyperpolarization that was approximately ten times greater than that predicted by the Nernst or constant-field equations. Electrogenic activity of the Na–K exchange pump was thought to be responsible for this high sensitivity to temperature. At temperatures below 5° C. the relationship between membrane potential and external K concentration could be predicted from the constant-field equation

$$V = \frac{RT}{F} \ln \frac{[K]_o + (P_{Na}/P_K)[Na]_o}{[K]_i}.$$

At $K_i = 235$ mM and $P_{Na}/P_K = 0.028$, a straight line could be obtained for changes in K_o at 4° C.

At temperatures above 5° C. the membrane potential could not be predicted from the constant-field equation except after inhibition of the electrogenic Na pump with ouabain or by reduction of the internal Na concentration. Inhibition of the electrogenic Na pump by low external K concentration was dependent on the external Na concentration. The membrane potential of the nerve-cell is thus the sum of ionic and metabolic components and the type of behaviour of the ionic component can be predicted by the constant-field type of equation.

Gorman and Marmor (1970b) investigated the effect of temperature changes on the Na-K permeability ratio. They found that the P_{Na}/P_K ratio increased from 0.028 at 4° C. to 0.068 at 18° C. This two to three times increase was primarily due to change in P_{Na}. The steady Na leakage into the cell is several times greater in the warm than in the cold and this may be the explanation of why the membrane potential of the neuron is more negative in preparations kept at the warmer temperature; the Na leakage may stimulate the electrogenic sodium pump, thus providing a mechanism to maintain the large steady-state potential at a warmer temperature.

SENSITIVITY TO OXYGEN TENSION

Work on a recognizable cell in the snail brain showed that the electrogenic Na pump in this preparation is sensitive to changes in Po_2 (Kerkut and York, 1969). The response of the cell to Po_2 changes during KAc injection is shown in *Fig.* 50. Lowering the Po_2 from a normal level of 100 mm. HgO_2 down to 70 mm. Hg reduced the membrane potential from 45 to 42 mV. When the Po_2 was increased to 250 mm. Hg the potential rose to 48 mV. During NaAc injection into this cell there was a large ouabain-sensitive hyperpolarization similar to that found by Kerkut and Thomas (1965). The sensitivity of the cell-membrane potential to changes in Po_2 was nearly doubled when the cell was hyperpolarized by Na injection. While the average change in membrane potential during KAc injection to a change in P_{O_2} of 250 mm. Hg O_2 was 4·6 ± 0·31 mV., the average change during NaAc injection was 8·6 ±0·6 mV. (*Table VI*). The sensitivity of the mem-

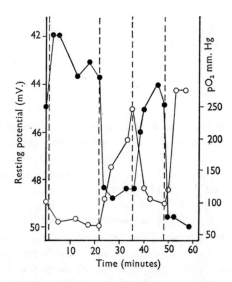

Fig. 50.—Effect of P_{O_2} on the membrane potential. Lowering the P_{O_2} decreased the membrane potential and raising the P_{O_2} increased the membrane potential. ●, P_{O_2}. ○, membrane potential. (*Figs.* 50, 51 reproduced from Kerkut, G. A., and York, B. (1969), '*Comp. Biochem. Physiol.*', **28**, 1125–1134.)

brane potential to P_{O_2} changes was reduced or even removed by 5×20^{-5} g. per ml. ouabain, as shown in *Fig.* 51. These results

Table VI.—SENSITIVITY OF SNAIL NEURON TO P_{O_2} CHANGES*

CHANGE IN MEMBRANE POTENTIAL	Na ACETATE	K ACETATE
Greatest change	22·5 mV.	7·5 mV.
Average change	8·6±0·6 mV. (57 values)	4·6±0·31 mV. (58 values)

* After Kerkut and York (1969)

suggest that the major part of the membrane potential response to P_{O_2} changes was due to an effect on the electrogenic Na pump.

The pump could be stopped by lowering the Po_2 and started again by increasing the Po_2. It could be that the pump sensitivity to Po_2 changes was due to the availability of only a small store of available high-energy compounds in the system. The great sensitivity of this preparation to Po_2 may be due to an inefficient intracellular respiratory pigment.

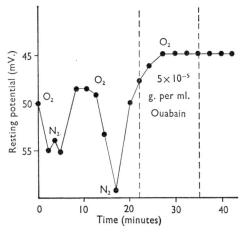

Fig. 51.—The effect of Po_2 and prolonged treatment with ouabain. Prolonged treatment with ouabain makes the neuron lose its oxygen sensitivity.

It is possible that the sensitivity of the membrane potential of spinal motoneurons to anoxia (Lloyd, 1953; Kolmodin and Skoglund, 1959; Eccles, Løyning, and Oshima, 1966) is partially due to an inhibition of an electrogenic Na pump present in these cells (Coombs, Eccles, and Fatt, 1955a,b; Ito and Oshima, 1965).

MEASUREMENT OF PUMP CURRENT

The electrogenic Na pump in snail neurons has been studied in more detail by Thomas (1968, 1969). Known quantities of Na were injected into the neurons by passing a current between two intracellular electrodes (Thomas, 1969). The membrane potential was measured through a third electrode. The current produced by the Na pump was measured by means of a feedback amplifier, whose output was fed into the cell through a fourth electrode.

As previously described, injection of Na into a neuron caused an increase in membrane potential which was abolished by ouabain and K-free Ringer. *Fig.* 52 shows that, while the first injection of

NaAc caused a hyperpolarization of 15 mV., ouabain reduced the hyperpolarization to 3 mV. after a second NaAc injection. K-free Ringer also reduced the Na-induced hyperpolarization from 12 to

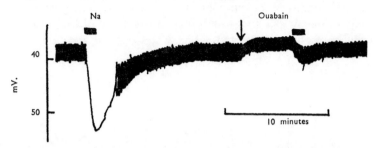

Fig. 52.—The response of the membrane potential of a snail neuron to Na injection before and after addition of 2×10^{-5} (w/v) ouabain. The 15-mV. hyperpolarization in response to Na injection was reduced to only 3 mV. by the ouabain. (*Figs. 52–57 reproduced from Thomas, R. C.* (1969), '*J. Physiol., Lond.*', **201**, 495–514.)

1·5 mV. However, when the external K was replaced 2 minutes later, there was an immediate further hyperpolarization of 15 mV. (*Fig. 53*). This result is similar to that obtained by Rang and Ritchie (1968) in work on post-tetanic hyperpolarization in nerve axons.

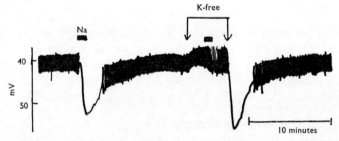

Fig. 53.—Effect of K-free Ringer on the response of the membrane potential of a snail neuron to Na injection. K-free Ringer reduced the hyperpolarization in response to Na from 12 to 2 mV. Readmission of the K 2 minutes later caused an immediate further hyperpolarization of 15 mV.

A measurement of the current generated by the Na pump was carried out using a 'slow clamp circuit' as snail neurons show a decrease in membrane resistance with increasing membrane potential. *Fig.* 54 shows the response of a snail neuron to Na injection before and after clamping. In *Fig.* 54 the membrane potential was held constant and a measurement of clamp current was obtained. This would be equal and opposite to that generated by the pump.

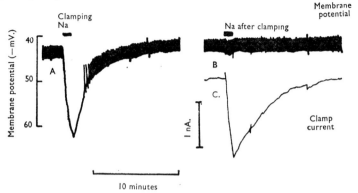

Fig. 54.—Response of snail neurons to Na injection before and after clamping. The clamp current is equal and opposite to that generated by the pump.

Measurements of changes in intracellular Na during NaAc injection were carried out using Na-sensitive electrodes. *Fig.* 55 A shows the Na-sensitive electrode designed by Hinke (1961). Unfortunately this design was not suitable for these experiments since when

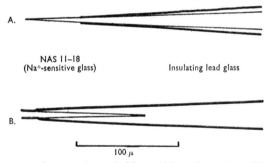

Fig. 55.—Construction of Na-sensitive electrodes. With the electrodes used by Thomas, it is possible to expose over 100 μ of Na-sensitive glass to the cell interior without having an electrode tip greater than 10 μ. A, Design used by Hinke (1961). B, Design used by Thomas (1969).

sufficient Na-sensitive glass was exposed beyond the insulation the electrodes were very difficult to insert fully into the neurons. Thus a second type of electrode was devised and constructed by Thomas, as shown in *Fig.* 55 B. This has the advantage that the insulation extends to the tip of the electrode so that the beginning of the insulation is always inserted into the cell.

Fig. 56 shows changes in internal Na and accompanying changes in clamp current during NaAc injection. The figure

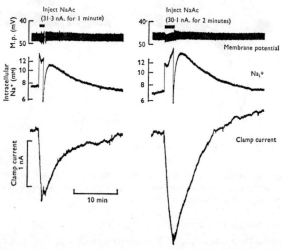

Fig. 56.—Response of the membrane potential, internal Na concentration, and clamp current to 2 injections of NaAc in snail neurons. The internal Na rises and falls at a similar rate to the clamp current.

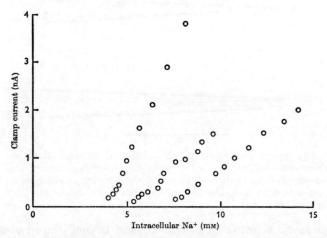

Fig. 57.—Relation between clamp current and internal Na, for injection of NaAc into 3 different snail neurons. The linear relationship shows that the clamp current is directly proportional to the internal Na concentration above the resting level.

shows how the internal Na concentration rises during Na injection and then declines towards the pre-injection level at a rate similar to that for the decline of the clamp current. From these experiments Thomas showed that the intensity of current generated was proportional to the excess of internal Na above the resting level as shown in *Fig.* 57, and that the total charge transferred was equivalent to the charge carried by between one-third and one-quarter of the Na extruded.

These very elegant experiments showed that about one-third of the Na extruded produces the electrogenic potential, the remaining two-thirds being coupled to the active transport of other ions, probably the uptake of K.

THE ELECTROGENIC SODIUM PUMP IN *Aplysia* NEURONS
TEMPERATURE SENSITIVITY

A large electrogenic Na pump was shown to be present in certain neurons of *Aplysia* (Carpenter and Alving, 1968). The

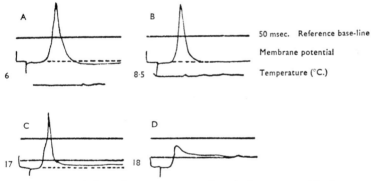

Fig. 58.—Effect of temperature on the membrane potential of an *Aplysia* neuron. As the temperature increased, the membrane potential hyperpolarized. (*Figs.* 58–62 *reproduced from* Carpenter, D. O., *and* Alving, B. O. (1968), '*J. gen. Physiol.*', **52**, 1–21.)

membrane potential of some neurons was very temperature-dependent and in some cells it increased with increasing temperature by up to 2 mV. per 1° C. Similar results were obtained earlier in *Aplysia* neurons by Carpenter (1967). The membrane potential was shown to exceed the E_K at room temperature in some cases, as determined by measurements of the equilibrium point of the spike after-potential. *Fig.* 58 shows recordings of the membrane potential and antidromic action potential from a giant cell at four different temperatures. At 6° C. the membrane potential was −48mV. and the after-potential was clearly more hyperpolarized than the membrane potential. On warming the membrane potential increased and was 56 mV. at 17° C. The after-potential was then more positive than the membrane potential. Further warming resulted in failure of generation of a soma spike, presumably because the hyperpolarization in the cell-body was so great.

CHLORIDE CONTRIBUTION AND CONDUCTANCE CHANGES

The hyperpolarization on warming was not due to an increased membrane permeability to Cl or to a Cl pump since replacement of NaCl by Na propionate caused no change in the magnitude of the membrane potential shift in response to a temperature change.

A comparison of *I–V* curves from a temperature-sensitive cell at four different temperatures showed that the hyperpolarization on warming could not be due to a conductance change to any ion.

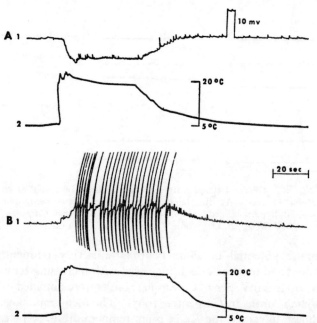

Fig. 59.—Effect of ouabain on the temperature response of *Aplysia* neurons. On warming 4×10^{-4} *M* ouabain caused the hyperpolarizing response to become a depolarization. A, Control. B, After addition of 4×10^{-4} *M* ouabain.

OUABAIN

Carpenter and Alving suggested that the hyperpolarization associated with increases in temperature could be the result of acceleration of active transport of Na. To test this, the membrane potential was compared in normal sea water and in sea water containing ouabain. The result of such an experiment is shown in *Fig.* 59. In AI the membrane potential at 5° C. was −45 mV., but on sudden warning it rapidly increased to 54 mV. However, after 8 minutes in 4×10^{-4} M ouabain, warming now produced an increase in synaptic activity and a depolarization (BI). The effect of ouabain on the response of the membrane potential to temperature changes was studied in over 35 cells. In every case in which hyperpolarization on warming was observed, it was markedly reduced or abolished within 2–5 minutes of exposure to 4×10^{-4} ouabain.

LITHIUM

Further evidence that the temperature sensitivity is a result of active transport was that the sensitivity was removed by lithium Ringer. As shown in *Fig.* 60, slow cooling in the presence

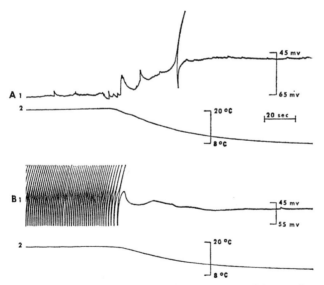

Fig. 60.—Effect of Li on the temperature response of the membrane potential in *Aplysia* neurons. A, In normal sea water. B, In sea water in which all the NaCl has been replaced by LiCl. Li removes the depolarization of the membrane potential in response to cooling.

of external Na caused a marked depolarization. After 20-30 minutes in sea water in which all NaCl had been replaced with LiCl, the cell discharged rapidly and on cooling the discharge ceased without a trace of depolarization.

K-FREE RINGER AND SHORT-CIRCUITING BY POTASSIUM

Removal of external K abolished or even reversed the hyperpolarization caused by warming, as shown in *Fig.* 61. In normal

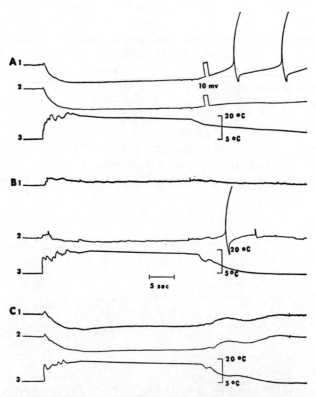

Fig. 61.—Effect of K-free sea water on the response of two *Aplysia* neurons (1 and 2) to temperature (3). A, In normal sea water. B, In K-free sea water. C, Less than 5 minutes after return to normal sea water. In K-free sea water the hyperpolarization on warming is abolished or even reversed.

sea water a rapid temperature change caused a hyperpolarization of 20 mV. which was maximal within less than 4 seconds. However, after 45 minutes in K-free sea water a similar temperature shift caused no hyperpolarization (*Fig.* 61 B). The effect was readily reversible on exposure to K (*Fig.* 61 C).

Cocaine, which increases the membrane resistance to passive K movements and would therefore reduce short-circuiting of the

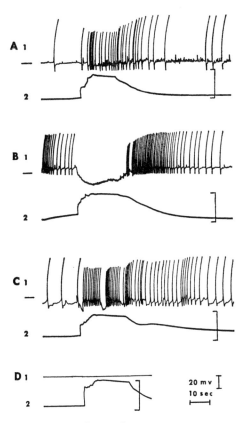

Fig. 62.—Effect of cocaine on the temperature response of *Aplysia* neurons. (A) In control experiments· B, During administration of $3 \times 10^{-3} M$ cocaine. C, After exposure to cocaine. D, Effect of temperature change on the electrode.

electrogenic Na pump by K, increased the membrane potential response to a rise in temperature. It was particularly effective in cells where there was only a small membrane potential shift initially as shown in *Fig.* 62. A temperature rise normally led to an acceleration of discharge which nearly obscured any hyperpolarization (*Fig.* 62 A). In the presence of $3 \times 10^{-3} M$ cocaine warming produced a clear-cut hyperpolarization of over 20 mV. (*Fig.* 57 B). The effect was reversed by washing (*Fig.* 62 C).

A rise in temperature did not affect the electrode when withdrawn from the cell, which showed that there was no D.C. shift with temperature in the recording equipment (*Fig.* 62 D).

The experiments where *Aplysia* neurons were kept at low temperatures for varying periods of time can be criticized since the neurons could have become loaded with Na under these conditions. However, Carpenter (1970) has subjected the neurons to sudden changes in temperature and found that cooling from 25 to 5° C. led to a decrease in the membrane potential by 35 mV., i.e., the potential dropped from −73 to −48 mV. This decrease in membrane potential was dependent upon Na and K and was abolished by ouabain.

In K-free sea water or in ouabain, sudden cooling often resulted in a hyperpolarization of up to 20 mV. This was interpreted in terms of a high P_{Na}/P_K ratio and a higher Q_{10} for P_{Na} than for P_K.

The electrogenic sodium pump could contribute as much as 50 mV. to the normal membrane potential. The membrane potential of the neurons of *Aplysia* so studied appears for all practical purposes to be set by the electrogenic sodium pump. In the absence of the pump the cell becomes rapidly depolarized, and there is later a further slow fall in the internal K concentration of the neuron which would also cause a reduction in the membrane potential.

The evidence described above points convincingly to the presence of an electrogenic Na pump in the membrane which contributes to the resting potential of *Aplysia* neurons. As in many other preparations this pump is dependent on the presence externally of K. The pump can be short-circuited by K in the absence of cocaine. The results exclude any explanation of temperature sensitivity in terms of E_K since the membrane potential can exceed the E_K as measured by the equilibrium point of the spike after-potential. Also it seems improbable that E_K is involved since a reduction of P_K by cocaine would decrease the contribution of E_K to the membrane potential, yet cocaine increased the response of the membrane potential to a rise in temperature.

THE ELECTROGENIC SODIUM PUMP IN STRETCH RECEPTOR NEURONS OF CRAYFISH

EFFECT OF HYPERPOLARIZING CURRENT

Evidence for an electrogenic Na pump was also found in the stretch receptor neuron of crayfish (Nakajima and Takahashi, 1965, 1966a, b). Firing activity in this neuron is followed by two types of after-potential. The first type, which follows a few spikes, is of short duration. The second type, which follows a train of spikes, is of long duration and has been termed 'post-tetanic hyperpolarization'. The properties of the short after-potential are explicable in terms of a selective increase in ionic permeability. This is not so for the post-tetanic hyperpolarization.

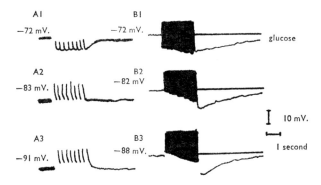

Fig. 63.—Effects of conditioning hyperpolarization on after-potentials in crayfish stretch receptor neurons. Hyperpolarization reversed the short after-potential but enhanced the post-tetanic hyperpolarization. (*Figs. 63–68 reproduced from Nakajima, S., and Takahashi, K.* (1966b), '*J. Physiol., Lond.*', **187**, 105–127.)

Intracellularly applied hyperpolarizing currents acted on the two types of after-potential in quite different ways, as shown in *Fig.* 63. The short after-potential was reduced (*Fig.* 63 A2) and eventually reversed (*Fig.* 63 A3) by a hyperpolarizing current. However, the post-tetanic hyperpolarization was slightly *increased* by conditioning hyperpolarization (*Fig.* 63 B3).

LITHIUM IONS AND 2,4 DINITROPHENOL

On replacing all the Na in the bathing Ringer by Li, the post-tetanic hyperpolarization disappeared. *Fig.* 64 A1, B1 shows the control hyperpolarization in normal solution. On replacing the solution with Li Ringer, the post-tetanic hyperpolarization disappeared in 5 minutes (*Fig.* 64 A2, B2). *Fig.* 64 A3, B3 shows that the effect of Li is reversed on reintroducing normal Ringer.

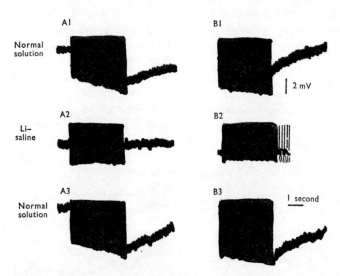

Fig. 64.—Effect of replacing Na with Li on the post-tetanic hyperpolarization in stretch receptor neurons of crayfish. Li reversibly abolishes post-tetanic hyperpolarization.

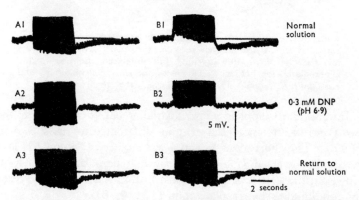

Fig. 65.—Effect of dinitrophenol on post-tetanic hyperpolarization in stretch receptor neuron of crayfish. Dinitrophenol removed the post-tetanic hyperpolarization.

After tetanus the post-tetanic hyperpolarization shown in *Fig.* 65 A1, B1 was about 3 mV. After application of 0·3 mM

2,4 dinitrophenol, it disappeared (*Fig.* 65 A2, B2). The effect was reversed on washing (*Fig.* 65 A3, B3). The inhibition of post-tetanic hyperpolarization by lithium ions and 2,4-dinitrophenol suggests that active Na efflux gives rise to the potential.

SODIUM INJECTION

A long-lasting hyperpolarization similar to post-tetanic hyperpolarization was brought about by electrophoretic application of Na into the cell. *Fig.* 66 shows the hyperpolarization obtained in the presence of tetrodotoxin (10^{-7} g. per ml.), a selective

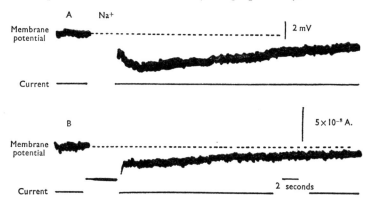

Fig. 66.—Electrophoretic Na injection into a crayfish stretch receptor neuron. The membrane potential hyperpolarized immediately after Na injection.

suppressor of spike generation in various excitable membranes. The magnitude of this hyperpolarization could be as large as 20 mV. This effect is similar to that observed by Kerkut and Thomas (1965) in snail neurons.

RELATION OF THE MEMBRANE POTENTIAL TO E_K

To show that the hyperpolarization was not due to the activity of an electroneutral pump causing a reduction of the external K concentration and a rise in E_K, the E_K was compared before and after tetanus. The E_K was measured by the size of the trough of the short after-potential, which is equal to E_K. *Fig.* 67 shows the short after-potential before (A) and after (B) tetanus. Although the membrane potential was enhanced 5 mV. by tetanus, the potential at the trough of the short after-potential was changed by only 1 mV. (*Fig.* 67 B).

This suggests that E_K was unchanged by tetanus. To support this, the relationship between the membrane potential and the

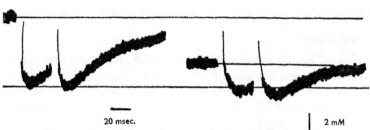

Fig. 67.—Effect of post-tetanic hyperpolarization on the trough level of the short after-potential in crayfish stretch receptor neuron. The trough level, which is a measure of E_K, was relatively unchanged after tetanus.

size of the short after-potential was plotted as shown in *Fig.* 68. Open symbols represent the membrane potential displaced by

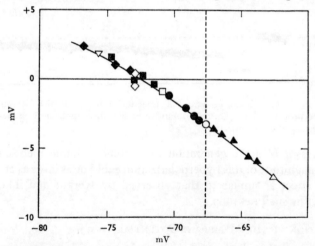

Fig. 68.—Relation of the membrane potential of a stretch receptor neuron to the size of the short after-potential. Open symbols: Membrane potential displaced by current. Closed symbols: Membrane potential displaced by post-tetanic hyperpolarization. The reversal potential of the short after-potential was not changed after tetanus.

externally applied currents. Closed symbols represent the membrane potential displaced by post-tetanic hyperpolarization. It can be seen that the reversal potential of the short after-potential did not change during post-tetanic hyperpolarization.

Nakajima and Takahashi concluded that the hyperpolarization was not caused by changes in external K but was probably due to the increase in activity of an electrogenic Na pump, stimulated by Na-loading of the cells during activity.

FURTHER EXPERIMENTS WITH LITHIUM

Obara and Grundfest (1968) also suggested that the stretch receptor neuron of crayfish possesses an electrogenic Na pump. Recordings from both the cell-body and the axon showed that Ringer in which Na has been replaced by Li caused considerable depolarization of the cell-body but not of the axon. After the preparation had been bathed in Li saline for 24 minutes, the soma had depolarized by about 16 mV., while the depolarization at the axonal recording site was only 4 mV. Obara and Grundfest suggest that the depolarization of the soma is caused by the interference of a Na pump system by Li. Only preliminary experiments have been carried out with drugs. Ouabain caused a gradual depolarization of a few millivolts with a time course similar to that of the Li-induced depolarization.

THE ELECTROGENIC SODIUM PUMP IN FROG SPINAL CORD GANGLIA

EFFECT OF BARIUM

An interesting paper by Nishi and Soeda (1964) describes the hyperpolarization of frog spinal cord ganglion cells by Ba. By

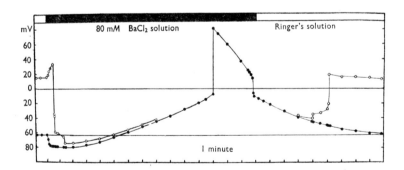

Fig. 69.—Hyperpolarization of a toad spinal ganglion cell by Ba Ringer. Addition of Ba caused an initial hyperpolarization of about 20 mV. ●, RP; ○, AP. (*Figs.* 69–71 *reproduced from Nishi, S., and Soeda, H.* (1964), '*Nature, Lond.*', **204**, 761–764.)

substituting isotonic $BaCl_2$ Ringer for normal Ringer, the membrane potential initially increased by about 20 mV. in less than 12 seconds.

This initial rise in membrane potential, which is shown in Fig. 69 A, was unaffected by Cl-free Ringer or by an increase in external K concentration. It was also independent of external Na. It was concluded that the equilibrium potentials of ions across the cell membrane were not involved in the Ba-hyperpolarization.

TEMPERATURE AND K-FREE RINGER

The most probable explanation for the hyperpolarization was that active extrusion of Na was responsible. This idea was confirmed by various experiments. For example, the hyperpolarization was shown to be accelerated by preincubation in K-rich Ringer and decreased by K-free Ringer.

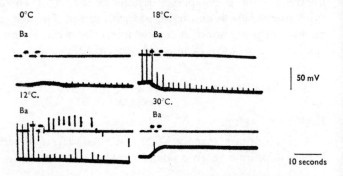

Fig. 70.—Effect of temperature on the hyperpolarization caused by Ba perfusion in toad spinal ganglion cells. The hyperpolarization was abolished by low temperatures (0–12° C.) and was reversed by high ones (30° C.).

The membrane potential during the first minute of Ba perfusion was very temperature-sensitive. Measurements of the membrane potential were taken after perfusion of the ganglion cells in different temperature Ringers (*Fig.* 70). The cells were perfused with normal Ringer solution at the same temperature as the Ba solution for more than 10 minutes before application of the Ba solution.

Ba solution was substituted for Ringer solution at the beginning of the first rectangular pulse on the zero potential line. Antidromic responses were evoked every 2 seconds except at 30° C.

At about 10° C. the hyperpolarization was present but was less than 5 mV. in size. However, between 18 and 22° C. a large hyperpolarization of about 20 mV. was seen. When the temperature of the Ba solution was raised to 30° C. or more, the membrane potential immediately depolarized and there was no sign of hyperpolarization. The Q_{10} of the hyperpolarization was about 3·0 over the range of 0–20° C. This dependency on temperature indicated that some chemical process was involved in the Ba hyperpolarization. The reduced response at high temperatures may be due to membrane damage.

Pre-incubation of cells with 1 mM 2,4-dinitrophenol always caused a complete disappearance of hyperpolarization at 20° C. This result again supports the idea that active Na transport is responsible for the hyperpolarization.

SODIUM INJECTION

Fig. 71 shows the effect of Ba solution of the membrane potential after electrophoretic injection of 3 M NaCl for 1–2 minutes.

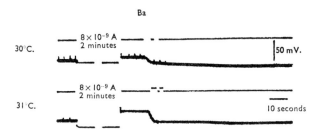

Fig. 71.—Effect of Na injection on the hyperpolarization caused by Ba perfusion in toad spinal ganglion cells. Perfusion of Ba after 1–2 minutes of iontophoretic Na injection increased the membrane potential by 30–40 mV.

The Na injection depolarized the cell by 5–40 mV. Applying Ba solution at the same temperature caused an increase in membrane potential of 30 or even 40 mV.

Nishi and Soeda concluded from these results that Ba caused a hyperpolarization by reducing P_K which would prevent coupling of the Na–K pump. The suppression of K movement would reduce short-circuiting and increase the potential generated by Na movement.

The Electrogenic Pump and the Parabolic Burster Cell

Strumwasser (1965) has studied the properties of a recognizable neuron of *Aplysia* which he terms the 'parabolic burster'. This cell produces bursts of action potentials over short periods of time, separated by quiet periods. The duration of the quiet periods or post-burst hyperpolarizations (POBH) is an exponential function of the number of spikes preceding the burst (*Fig. 72*).

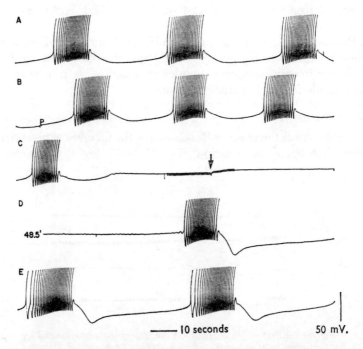

Fig. 72.—Effect of KCl injection on the parabolic burster cell of *Aplysia*. Injection begins at P (line B) and finishes at arrow (line C). KCl injection suppresses the bursts and leads to prolonged interburst interval time. (*Reproduced from Strumwasser, F. (1965) in 'Circadian Clocks' (ed. Aschoff, J.). North-Holland: Amsterdam.*)

The bursting activity can be abolished by injection of KCl (200 picolitres per 250 μ diameter cell) but not by K_2SO_4 (*Fig. 72*). When the process of burst generation returns several minutes after KCl injection, the POBH is very much increased and the trough is greater than the undershoot of the spikes during the burst (*Fig. 72 D, E*). Reducing the external Cl concentration to one-quarter of the normal level abolished the POBH. These results suggest that POBH is due to the initiation of a specific increase in Cl conductance near the end of the burst. The entering Cl could be removed by an outwardly directed Cl pump.

However, further experiments suggest that the electrogenic Na pump, shown to be present in *Aplysia* neurons (Carpenter and Alving, 1968), is also involved in the POBH. The POBH initiated by a depolarizing pulse can be blocked by ouabain. Substitution of Li for Na in the bathing sea water also blocks POBH. Strumwasser suggested that in this cell, which maintains both a low internal Cl and Na, the POBH could be generated by an electrogenic Na–Cl loosely linked pump.

As the POBH can still occur in the presence of tetrodotoxin it seems improbable that the electrogenic Na–Cl pump is stimulated by an increase in internal Na during firing activity, since tetrodotoxin blocks the Na entry. It is possible that the pump rate could be controlled by the internal Cl concentration instead.

An Electrogenic Pump in *Neurospora*

Very good evidence was reported by Slayman (1965) for active ion movements contributing 85 per cent of the membrane potential of the fungus, *Neurospora crassa*. Unfortunately the ions involved in the system are not known and are probably not Na, as Na extrusion from this cell is weak and the membrane potential is not sensitive to ouabain. However, anoxia and a wide variety of metabolic inhibitors reduced the internal potential of the cell from about -200 mV. to -30 mV.

The decline of the membrane potential on addition of respiratory inhibitors was immediate and rapid. *Fig.* 73 shows the effect of 1 mM Na azide on the membrane potential. The initial rate of change of membrane potential was more than 20 mV. per second. When 10 mM azide was added the initial rate was increased to 20 mV. per second. Within a minute the internal potential had fallen by more than 200 mV. When the azide was removed the potential recovered gradually over 8–10 minutes.

Fig. 74 shows the effect of 2,4-dinitrophenol, carbon monoxide, nitrogen, and azide on the membrane potential. Carbon monoxide and nitrogen both reduced the internal potential from about 200 to 100 mV. Both dinitrophenol and azide had a greater effect.

The total membrane resistance fell very little during the periods of rapid potential change. Since there is no resistance shift it appears that the change in membrane potential is not due to an ionic diffusion potential. A depletion mechanism similar to that described in nerve and muscle was discounted on the grounds that the cell wall of *Neurospora* is an insubstantial barrier to small ions. A streaming potential due to an osmotic gradient would give rise to a potential of the opposite polarity to that noted. Slayman

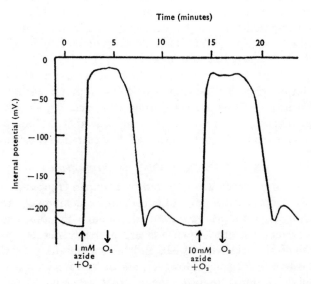

Fig. 73.—Response of the internal potential of *Neurospora* to azide. Azide reduced the internal potential from -240 mV. to -40 mV. within a minute. (*Figs.* 73, 74 *reproduced from* Slayman, C. L. (1965), '*J. gen. Physiol.*', **49**, 93–116.)

concluded that the large change in potential was due to an electrogenic process probably involving H^+ rather than Na^+. This type of potential generation could be similar to the process suggested by Conway, which he described in his 'redox theory' (Conway, 1964).

The existence of cytochrome-like enzymes in the cell membrane which could transfer the H^+ has not yet been demonstrated in any species but bacteria.

Slayman (1970) suggested that the electrical potentials in *Nitella* and *Neurospora* may be explained as being due to two components.

The first arises from the diffusion of Na^+, K^+, and H^+ down their electrochemical gradients. The second potential component is

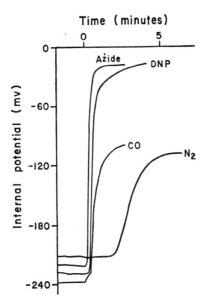

Fig. 74.—Response of the internal potential of *Neurospora* to 1 mM azide, 1 mM dinitrophenol, carbon monoxide, and anoxia. All these inhibitors reduced the internal potential by 100–200 mV.

due to the utilization of energy and the active efflux of hydrogen ions associated with an electrogenic H^+ pump.

PART IV

THE ELECTROGENIC SODIUM PUMP IN MUSCLE-FIBRES

PROBLEMS OF INTERPRETATION

THE method used by most workers to show the presence of an electrogenic Na pump in muscle involves Na-loading the fibres by soaking for varying lengths of time in cold K-free Ringer. The measured membrane potential in a K-containing recovery medium is compared with the value of E_K, calculated from the levels of K found in the muscle during the K influx and Na efflux. The membrane potential measured during the initial recovery period often exceeds the calculated E_K, but is decreased by ouabain, low temperature, and metabolic inhibitors. However, attempts to show that the increase in membrane potential beyond E_K during recovery is caused by stimulation of an electrogenic Na pump have been complicated by a number of factors.

First, it is very difficult to correlate changes in membrane potential measured in surface cells with E_K estimated by measurement of K concentrations of the whole muscle, since diffusion of K to the interior of the muscle during recovery takes some time, while K uptake in the surface cells, whose membrane potential is monitored, will be relatively rapid. Therefore the membrane potential of a surface cell might exceed the average E_K of the whole muscle, but could still be lower than the E_K of the muscle.

A second difficulty is that, as in nerve, the external K may be reduced immediately exterior to the muscle-cells by an electroneutral pump during recovery because of diffusion barriers. This would occur if the K was taken up into the muscle-cells at a faster rate than it could be replaced from the bulk of the solution. Some of the hyperpolarization during recovery could then be explained as an increase in E_K due to a fall in external K.

Another problem is that errors in the determination of E_K can arise from incorrect assumptions of the extracellular space or from changes in the fibre volume which will change the ion concentration.

Discovery of the Electrogenic Sodium Pump in Frog Muscle Net Sodium Loss

Carey, Conway, and Kernan (1959) studied frog sartorius muscles to find the conditions required for Na-loading the fibres and for obtaining a net Na loss from the fibres on recovery by the Na pump. This work showed that after muscles were Na-loaded by soaking in cold K-free Ringer containing 120 mM Na overnight, they must be soaked in a 10 mM K recovery medium containing a lower Na concentration of about 104 mM to obtain a *net* Na loss on recovery.

If the muscle was soaked in 104 mM Na or placed in a recovery medium containing 120 mM Na, then there was little or no average Na excretion. Conway, Kernan, and Zadunaisky (1961) decided that the need to reduce Na in the recovery medium was due to the presence of a critical energy barrier to Na excretion consisting of osmotic and electrical gradients. Kernan (1962b) postulated that the level of this barrier was determined by the available energy in the cell extruding Na. Addition of lactate or insulin to the bathing Ringer solution caused the cell to excrete Na under conditions where this would not normally have occurred.

Relation of the Membrane Potential to E_K

The first demonstration of an electrogenic Na pump in Na-loaded muscles was carried out by Kernan (1962a, b). In preliminary experiments frog sartorii were Na-loaded overnight in cold K-free Ringer (120 mM Na) and then the membrane potential was recorded on immersion in 10 mM K recovery medium. Active Na extrusion was facilitated by immersion in Na-free choline Cl Ringer. The value of the E_K was calculated from muscle K levels estimated by flame photometry. During Na extrusion into choline Cl the measured membrane potential exceeded the calculated potential by 33·7 mV. During extrusion into 104 mM Na-containing solution, the observed membrane potential exceeded the calculated potential by 11·1 mV. These results are shown in *Table VII*.

Similar results were obtained in Cl-free Ringer (*Table VIII*), thus eliminating any contribution of Cl to the difference in measured and calculated membrane potential.

In later experiments (Kernan, 1962b) membrane potentials of Na-loaded frog sartorii were measured during Na extrusion in a recovery medium containing 120 mM Na, 1 mM lactate, and 30 U. per litre insulin. After 5 minutes of immersion in this solution membrane potentials were recorded for 15 minutes. The

Table VII.—Observed and Calculated Membrane Potentials of Sodium-rich Frog Sartorii before and after Extrusion of Sodium into Recovery Fluids*

Experiment	Mean Membrane Potential (mV.)	
	Observed	Calculated
In choline chloride		
Before extrusion	83·8±0·6	50·1±2·6
After extrusion	60·5±0·6	58·0±2·1
120/104 sodium procedure		
Before extrusion	63·4±0·7	52·3±0·9
After extrusion	56·0±1·1	58·0±0·6

* After Kernan (1962a).

Table VIII.—Observed and Calculated Mean Membrane Potentials of Frog Sartorii before and after Extrusion of Sodium into Chloride-free Recovery Fluids*

Experiment	Mean Membrane Potential (mV.)	
	Observed	Calculated
Sucrose–Ringer		
Before extrusion	75·5±0·7	60·1±1·1
After extrusion	—	—
Sulphate–Ringer		
Before extrusion	68·9±0·6	57·5±0·6
After extrusion	70·0±0·5	67·1±0·5

* After Kernan (1962a).

mean value of the membrane potential was found to be 97·6± 0·3 mV. From a measurement of muscle K levels and the extracellular space, the E_K was calculated and found to be 91·0± 0·8 mV. Thus the difference between the observed and calculated potentials was 6·6 mV.

Effect of o-Phenanthroline

When Na transport was inhibited by addition of the inhibitor o-phenanthroline, the membrane potential fell from 97 mV. to a mean value of 93·4±0·8 mV. There was then no significant difference between the membrane potential and E_K. Kernan concluded that extrusion of Na by an active process was responsible for the increased internal potential, the K entrance occurring passively to restore conditions of electrical neutrality within the fibres.

Removal of Short-circuiting by Chloride

Kernan and Tangney (1964) later produced some convincing evidence that the hyperpolarization during recovery of Na-loaded muscles was due to the activity of an electrogenic Na pump. They eliminated any contribution of Cl to the raised membrane potential by replacing the NaCl of the soaking and recovery media with Na-methyl sulphate. Under these conditions the muscle did not excrete Na, so the external K concentration of the recovery medium was raised to 25 mM. In this Cl-free solution the membrane potential exceeded the E_K by as much as 25 mV. This large difference in potential would be hard to explain in terms of a neutral pump. Kernan and Tangney suggested that in a Cl-containing solution Cl leaves the cell passively during active Na excretion and short-circuits the electrogenic Na pump. As found in nerve by Rang and Ritchie (1968), the replacement of external Cl by a large impermeable ion removes this short-circuiting by depleting the intracellular Cl.

Effect of Denervation

The effect of denervation of the frog skeletal muscle on the functioning of the electrogenic Na pump was then studied by Kernan (1966). The preparation used consisted of companion sartorii connected to the spinal cord of a decapitated frog by way of the sciatic nerves. The sartorii were Na-loaded overnight by immersion in cold K-free Ringer (120 mM Na). Na excretion occurred on transfer of the muscle to Ringer–Conway fluid containing 104 mM Na and 10 mM K. One muscle of each pair was cut free of the nerve before immersion in the recovery medium, and the Na excretion, K uptake, and membrane potential of innervated and denervated muscles was compared. *Table IX* shows the result of these experiments. The sodium pump becomes more electrogenic (i.e., generates a greater potential) after denervation of the muscle, while Na excretion is decreased. Kernan suggests that the increase in potential could be due to a reduced K uptake or permeability.

Table IX.—FINAL CONCENTRATIONS OF SODIUM IONS AND POTASSIUM IONS IN MUSCLES AFTER RECOVERY, AND E_m MEASURED 10 MINUTES AFTER RE-IMMERSION OF SODIUM-RICH MUSCLES AND 5 MINUTES AFTER DENERVATION*

MUSCLE	Na^+ (mEq. per kg.)	K^+ (mEq. per kg.)	E_m (mV.)
With nerve	33·1±2·9	87·9±2·5	63·6±0·9
Denervated	45·9±2·3	78·9±2·4	72·0±0·9
E_K		59·2±3·2	

* After Kernan (1966).

FURTHER WORK ON THE ELECTROGENIC SODIUM PUMP IN FROG MUSCLE

SUMMARY OF WORK

Keynes and Rybova (1963) confirmed Kernan's finding that during recovery of Na-loaded muscles the membrane potential exceeded the calculated E_K. They found a difference of about 15 mV. between the two values, which was abolished by ouabain, low temperature, and Li.

Since 1963, a number of studies on the membrane potential during active Na efflux from Na-loaded muscle have supported the theory that some or all of the outward movement of Na from muscle-cells is capable of generating a potential (Adrian and Slayman, 1964, 1966; Hashimoto, 1964; Cross, Keynes, and Rybova, 1965; Frumento, 1965; Mullins and Awad, 1965; Page and Storm, 1965; Harris and Ochs, 1966; Geduldig, 1968a, b; Taylor, Paton, and Daniel, 1969). The hyperpolarization during recovery is abolished by ouabain, cold, and replacement of external Na by Li. This suggests an involvement of active Na transport in the generation of the hyperpolarization. Very careful observations have been required to eliminate other factors involved in the change of membrane potential during recovery.

Removal of Potassium Short-circuiting

The criticism that an increase in membrane potential during recovery could be due to a depletion of K immediately exterior to the membrane by an *electroneutral* pump was overcome by the use of Ringer solutions in which K was replaced by Rb. Adrian and Slayman (1964) prepared Na-rich muscles by soaking frog sartorius muscles in K-free Ringer at 1° C. for 48 hours and then allowing the muscles to recover in solutions containing 10 mM Rb or 10 mM K at 18° C. The membrane potentials of the muscles were recorded in the recovery media before pumping, at 1° C., and after pumping had begun, at 18–20° C. The muscle membrane potentials increased to a maximum over a period of 10–20 minutes and then slowly declined. The mean internal potential in Rb recovery solution was -90 to -105 mV. as compared with -75 to -90 mV. in K recovery fluid.

If a *neutral* pump had been involved in Na extrusion the maximum membrane potential recorded in K Ringer would indicate a local reduction of K external to the cells from 10 mM to 2–5 mM. Since Rb is pumped nearly as fast as K the reduction of external Rb should be of the same order. However, results of Hodgkin and Horowicz (1959) showed that the membrane potential of resting muscles was smaller in Rb than in K at equal concentrations below 4 mM, and that the mean potential was only 3 mV. more negative in Rb at 5 mM. However, since the membrane potential in a Rb-recovery medium was 15 mV. greater than in a K-recovery medium, the hyperpolarization could not be solely due to K or Rb depletion by an electroneutral pump. From the measured pumping rate of Rb it seems unlikely that the pump is 100 per cent electrogenic. The increase in potential in the presence of Rb as compared with K is probably due to a reduction in short-circuiting of the electrogenic Na pump by passive inward K movement since the membrane is less permeable to Rb than to K.

Temperature and Recovery

Other evidence suggesting that not all the K influx in recovering Na-loaded muscles is chemically coupled to Na efflux was described in a later paper by Adrian and Slayman (1966). On changing the temperature of the Rb-recovery medium surrounding Na-loaded muscle-fibres from 4 to 32° C., the internal potential of the surface fibres showed a huge rise from −49 mV. to −109 mV., as shown in *Fig.* 75. The large size of this hyperpolarization is difficult to explain on the basis of depletion of external Rb by a neutral pump. Any contribution of Cl was excluded by depleting the muscle of Cl during Na loading and using SO_4^{2-}-containing recovery Ringer.

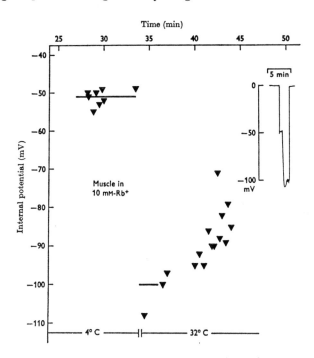

Fig. 75.—Effect of sudden warming on the internal potential of surface fibres of a Na-rich, Cl-depleted muscle. (Insert record shows the potential change which occurred during warming.) Warming from 4 to 32° C. caused a huge rise in membrane potential from 49 to 109 mV. (*Reproduced from Adrian, R. H., and Slayman, C. L.* (1966), '*J. Physiol., Lond.*', **184**, 970–1014.)

Cocaine, which has been shown to increase the membrane resistance to K (Adrian and Freygang, 1962), would decrease the contribution of E_K to the membrane potential during recovery. However, it would increase the potential generated by uncoupled Na efflux by decreasing the short-circuiting effect of the K inward movement driven by the generated potential difference. Adrian and Slayman found that the potential of Na-loaded muscles warmed in the absence of cocaine rose from 57 mV. to 70–75 mV. When 3 mM cocaine was added the average potential rose to about 90 mV. in about 10 minutes. Taking changes of E_K into account, the hyperpolarization produced by the pump was increased by a factor of 2–3. Membrane-resistance measurements also showed a two- to threefold increase.

Harris and Ochs (1966) obtained similar results in recovering Na-loaded muscles. When cocaine, procaine, amytal, or mepyramine was added to the recovery medium the rise in mean potential on warming was increased twofold, as shown in *Table X*, and the membrane resistance was also increased.

Table X.—Effect of Drugs on Maximum Potential Changes and on Ion Movements of Muscles warmed from 5 to 26° C. in Solutions with 5 mM KC added

Addition	Increase of Potential Difference (mV.)
None	8·0*
1 mM mepyramine	17·8
1 mM procaine	18·1
1 mM amytal	17·3
3 mM cocaine	17·6

* In similar experiments without drug addition the potential rise observed was 10·3, 10·6, and 14·7 mV.

Validity of E_K Measurements in Frog Sartorius

A careful study by Cross and others (1965) showed that the difference between the membrane potential and E_K of recovering Na-loaded muscles is not solely due to an incorrect assessment of the true E_K. Frog muscles were Na-loaded by soaking in K-free Ringer at 22° C. for various periods of time in the order of 24 hours, after which the membrane potential of the surface fibres was measured. E_K was calculated by analysis of the muscle K by flame photometry. *Fig.* 76 shows the change in membrane potential of the fibres of 9 different muscles during recovery in 10 mM K at room temperature after Na-loading. In K-free solution the membrane potential was over 100 mV. On raising the external K to 10 mM it fell to 71·5 mV. and remained there for about 45 minutes, after which it declined slowly to 63 mV. Analysis of companion muscles showed that the E_K (solid line on the graph) was 15–20 mV. smaller than the membrane potential during the first 30 minutes of recovery. The calculation of E_K depended on two assumptions. The first was that the extracellular space was no more than 130 ml. per kg., a value found by Desmedt (1953). However, if the estimated extracellular space was too small, this would make the calculated E_K value too large rather than too small. The second assumption was that the activity coefficients were the same inside and outside the fibres. Work by Lev (1964), using intracellular K-sensitive electrodes, suggested that there was no inequality in the activity coefficients on the two sides of the membrane.

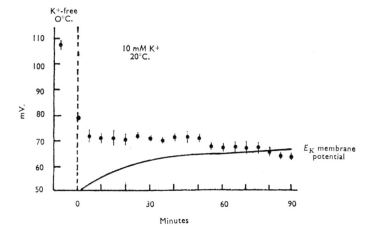

Fig. 76.—Average resting membrane potential in Na-loaded frog sartorius muscle fibres before and after transference to a 10 mM recovery solution at 20° C. The membrane potential exceeded the calculated E_K during the first hour of recovery. (*Figs. 76, 77 reproduced from Cross, S. B., Keynes, R. D., and Rybova, R. (1965), 'J. Physiol., Lond.', 181, 865-880.)

CHLORIDE CONTRIBUTION

Any effect of the passive distribution of Cl in maintaining the raised potential is unlikely to have persisted for one hour. However, Cross and others (1965) excluded the possibility that the high membrane potential was maintained by the Cl distribution by blocking active Na transport with $10^{-5} M$ ouabain. If the Cl distribution was responsible for the high membrane potential ouabain should have no effect on the potential during recovery. With ouabain in the recovery medium, measurements of the membrane potential after 5–10 minutes of recovery showed that with active Na transport blocked, the membrane potential was *smaller* than E_K by 3·1–1·2 mV. The values of the membrane potential and E_K 5–10 minutes and 60 minutes after recovery are shown in Table XI.

Table XI.—Membrane Potentials in Muscle treated with Recovery Solution containing 10 mM K and 10^{-5} M Ouabain

Soaking-in Time (hr)	E_m in Soaking-in Solution (mV.)	E_m (mV.)		E_K (mV.)		K Content (mmole per kg.)	
		Initial	Final	Initial	Final	Initial	Final
24*	31	35	36	47	51	45·2	53·3
24*	75	60	52	54	58	57·2	68·2
24*	30	34	28	53	55	54·9	59·9
24*	68	57	43	56	55	61·0	60·8
21	51	42	34	55	54	61·1	57·1
17	64	53	47	58	59	66·1	70·9
21	42	30	24	58	58	67·0	67·0
6½	46	45	42	58	60	67·3	75·4
4½	120	60	56	61	63	80·7	83·9
8	92	66	51	63	62	77·7	77·6
23	74	54	42	59	60	68·9	72·4
5	113	56	51	63	61	75·3	72·1
24	79	55	49	(58)	58	—	65·9
24	82	56	49	(60)	60	—	72·0

* In these experiments there was ouabain only in the recovery solution. In all the other experiments ouabain was also added to the soaking-in solution, at least 40 minutes before the end of the soaking-in period.
Temperature was about 2° C. during period of soaking-in and was 19–24° C. during recovery.

Effect of Temperature

Low temperature was also used to block active Na transport. Membrane potential determinations were made first at 3° C. in the soaking solution and then in the recovery solution containing 10 mM K also at 3° C., as shown in *Fig.* 77. After 20 minutes in the cold recovery solution the membrane potential had fallen to a value very close to E_K, calculated from analysis of the companion muscle. When the temperature was then raised to 19° C. the membrane potential rose rapidly by about 10 mV. (*Fig.* 77). However, the corresponding change in E_K was only 3·2 mV. (solid line in graph). It was concluded that the greater part of this hyperpolarization arose from acceleration of active Na transport on warming.

As Cross and others point out, the difference between the membrane potential and E_K in the experiment shown in *Fig.* 77 is unlikely to be solely due to a difference in E_K of the surface fibres and the cells in the interior of the muscle for the following reason: the surface cells will have a lower internal K concentration after soaking than the average for the muscle at zero time since they have been in direct contact with the K-free Ringer. Thus

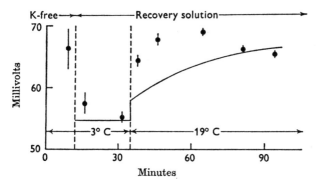

Fig. 77.—Change in membrane potential when active Na transport is initiated by raising the temperature from 3 to 19° C. (Solid line shows the calculated value of E_K.) The membrane potential exceeded the E_K during active Na efflux.

the E_K of the surface cells whose membrane potential is recorded will be *lower* than the average for the muscle, and so the difference between the membrane potential and the E_K would be greater than that calculated. However, after some minutes in 10 mM K, the uptake of K would be greatest in the surface fibres and the difference between the membrane potential and the E_K would be less than shown in the figure.

It is unlikely that the hyperpolarization could be due to a depletion of local external K just exterior to the surface cells by an electroneutral pump, as there is no apparent diffusion barrier for these cells from which the hyperpolarization was recorded.

The results obtained by Cross and others (1965) would be difficult to explain on an exclusively electrical or exclusively chemical coupling of the Na pump. Using a value of gK calculated by Hodgkin and Horowicz (1959) to determine the net K influx, the hyperpolarization could just be explained if all the Na efflux was electrogenic. However, as gK is variable the results still allow some chemical coupling with K. That coupling of K and Na can occur during recovery is certain, as the membrane potential was less than 40 mV. after soaking in cold K-free Ringer during certain months, so that the membrane potential was less than E_K. Since Na and K transport were not impaired, this showed that K could move inward against an electrochemical gradient.

Further evidence for electrogenic Na efflux in recovering Na-loaded muscles was demonstrated by Mullins and Awad (1965).

Table XII.—MEMBRANE POTENTIALS OF CONTROL AND Na-LOADED MUSCLES

	K-FREE SULPHATE RINGER		Li_2SO_4 RINGER'S, 20° C., 2·5 mM (K)	
	4° C. (mV.)	20° C. (mV.)	0·5–2·0 minutes (mV.)	90–130 minutes (mV.)
Membrane potential for muscles in sulphate Ringer 12 hours at 20° C. upon transfer	−104±2 (6)*	−107±2 (6)	−88±1 (6)	−82±2 (6)
E_K (calculated)			−95	−89
Membrane potential for muscles in K-free sulphate Ringer for 12 hours at 4° C. (Na-loaded) upon transfer	−47±5 (10)	−76±4 (10)	−98±4 (10)	−83±3 (10)
E_K (calculated)			−82	−88

* Variability is given as ±S.D. The number of muscles used is shown in parentheses.

The use of SO_4-containing solutions throughout the experiments removed any contribution of Cl to the membrane potential during recovery. Mullins and Awad compared the membrane potential of muscles equilibrated for 12 hours at 20° C. and at 4° C. (Na-loaded) when transferred to K-free Ringer at 4° C. and 20° C., and to Li_2SO_4 solution (2·5 mM K) at 20° C. *Table XII* shows the result of these experiments. Muscles with a normal internal Na concentration showed only a slight increase in membrane potential with a temperature rise from 4 to 20° C., compared with

Na-loaded muscles. The Na-loaded muscles showed a large increase in membrane potential, which gradually declined in Li Ringer. The membrane potential exceeded the E_K by 13 mV. 0·5–2 minutes after transfer. These results, like those of other authors described in this section, suggest that recovery on warming after Na-loading muscle-fibres leads to an increase in membrane potential generated by an electrogenic Na pump.

THE ELECTROGENIC SODIUM PUMP IN CAT HEART MUSCLE

Page and Storm (1965), working on cat heart muscle, obtained results similar to Cross and others (1965). Cat right papillary muscle was depleted of K and enriched with Na, Cl, and H_2O by

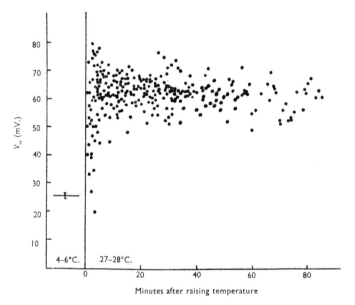

Fig. 78.—Effect of sudden rewarming on the membrane potential of Na-loaded cat heart muscle. The membrane potential recovery was complete within 1–9 minutes. It increased from 25·6 ± 0·7 mV. in the cold to 61·3 ± 0·5 mV. after 10 minutes' warming. (*Figs.* 78, 79 *reproduced from* Page, E., *and* Storm, S. R. (1965), '*J. gen. Physiol.*', **48**, 957–972.)

pre-incubation at 2–3° C. for 1–2 hours. The recovery process was studied after sudden rewarming to 27–28° C. As shown in *Fig.* 78, recovery of the membrane potential of the cells at the muscle surface was complete after 1–9 minutes. The mean value of the membrane potential in the cold was 25.6 ± 0.7 mV. as compared with an average of 61.3 ± 0.5 mV. at 10–85 minutes after rewarming. Recovery to a stable level was observed to occur in less than 1 minute in 4 experiments. In 2 experiments where the electrode remained in the cell during a change from cold to warm solutions, the membrane potential recovered in 15 seconds

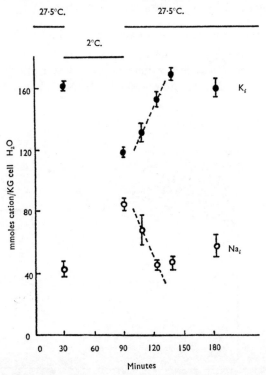

Fig. 79.—Effects of cooling and rewarming on the intracellular concentrations of Na and K in cat heart muscle. Restoration of the internal Na and K took at least 30 minutes.

in one case, and in a much smaller time in the other. However, only a slight restoration of cellular ion contents, measured in the whole muscle, had occurred in 10 minutes (*Fig.* 79). At physiological external K concentrations recovery of ion and water was complete within 30 minutes. 10^{-4} M ouabain completely prevented recovery of the membrane potential on warming.

Page and Storm (1965) suggested that the extrusion of Na and the uptake of K into the surface cells by an electroneutral pump might greatly exceed net ion movements found by chemical analysis of the whole muscle, since during recovery the K in the muscle interior could be taken up faster than it could diffuse in from the solution. However, they calculated that, if the recovery of the membrane potential of the surface cells occurred only by complete restoration of the internal Na and K concentrations to

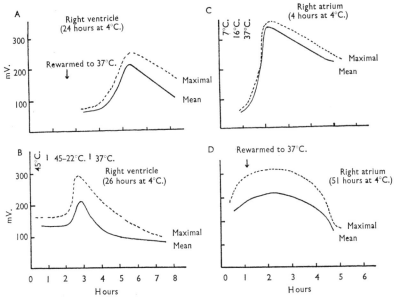

Fig. 80.—Resting potentials of right ventricular and atrial muscles during rewarming after prolonged hypothermia at 4° C. On warming the membrane potential showed a huge increase over several hours. The increase in potential was up to 180 mV. in the right ventricular muscle. (*Figs. 80–82 reproduced from Tamai, T., and Kagiyama, S.* (1968), '*Circulation Res.*', **22**, 423–433.)

normal, the rate of K influx required to lead to the rapid increase in membrane potential would have to be 180 times greater than the steady-state exchange. Therefore they concluded that an electrogenic Na pump was responsible for the rapid membrane potential change.

A huge membrane potential (267 mV.) was recorded by Tamai and Kagiyama (1968) from recovering Na-loaded cat heart muscle. The muscles were stored at 4° C. for varying periods of time, after which the right atrium or a strip of the wall of the right ventricle was dissected out and placed in cold Tyrode solution. This solution was gradually warmed until the muscle was at 37° C. After warming the membrane potential of the recovering muscles was monitored using intracellular recording. *Fig.* 80 shows the membrane potential of the right ventricular (*Fig.* 80 A, B) and atrial (*Fig.* 80 A, B) muscles during rewarming after prolonged hypothermia. Open circles represent results from silent cells, closed circles represent results from spontaneously active cells. The very big hyperpolarizations observed led Tamai and Kagiyama to consider that an electrogenic Na pump was responsible since the membrane potential was too high to be explained by E_K. *Table XIII* shows the membrane potential of ventricular and

Table XIII.—MEMBRANE POTENTIALS (mV.) OF VENTRICULAR AND ATRIAL MUSCLES OBTAINED AT A MAXIMALLY HYPERPOLARIZING STAGE DURING RECOVERY AT 37° C.*

Hours Preserved at 4° C.	Right Ventricle (11 cats)	Right Atrial Appendage (10 cats)	S-A and A-V Nodes (9 cats)
0	87·3±0·9 (16)	83·7±1·2 (34)	67·4±1·1 (19)
1·5	—	71·1±1·7 (19)	68·7±1·9 (14)
12·5	91·2±0·5 (8)	—	—
15	106·5±2·2 (17)	80·5±1·7 (15)	73·5±2·2 (9)
19	267·7±16·4 (10)	—	—
20	—	107·3±7·5 (10)	—
20·3	229·9±13·7 (17)	—	—
23·4	145·6±4·5 (21)	—	—
23·7	207·3±5·2 (23)	—	—
26	212·7±19·4 (10)	—	—
43	—	103·7±3·2 (8)	112·5±2·8 (6)
44·3	—	148·3±6·6 (15)	103·9±15·5 (4)
46	157·1±5·2 (9)	—	—
50	179·2±2·7 (14)	—	—
51	—	184·4±3·2 (18)	154·4±5·4 (13)
71	—	118·6±1·5 (16)	105·3±4·8 (6)
74	—	156·8±4·5 (11)	114·7±3·9 (8)
144	—	78·6±8·6 (17)	76·1±7·5 (14)
169	105·1±3·4 (11)	—	—

* After Tamai and Kagiyama (1968). Values are means ±S.E. Numbers in parentheses are number of measurements. Each line represents measurements made on a different cat, except for 0 and 15 hours when 2 cats were studied (1 for measurements on the ventricle and 1 on the atrium and the nodal tissues).

atrial muscles recorded at a maximally hyperpolarized stage during recovery at 37° C. Maximum hyperpolarizations were obtained in ventricular muscles that had been preserved for 20 hours at 4° C. and in atrial muscles that had been preserved at 50 hours at 4° C. Maxima were often not reached for 3 hours after rewarming.

Further evidence that this hyperpolarization was due to the activity of an electrogenic Na pump was that ouabain (10^{-5}–10^{-6} M) caused an immediate depolarization during recovery (*Fig.* 81).

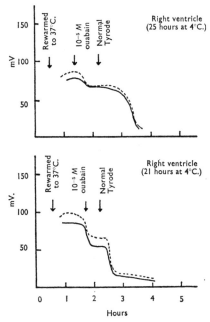

Fig. 81.—Effect of 10^{-5} and 10^{-6} M ouabain on the membrane potential of rewarmed right ventricles. Ouabain caused an immediate depolarization during recovery. ---, Maximum; ——, mean.

The lowered membrane potential did not recover when the preparation was returned to normal Tyrode solution.

The replacement of 90 per cent of the external Na by Li also interfered with the recovery process. After storing heart muscle in Li–Tyrode solution the membrane potential gradually declined over 1 hour. A return to normal Tyrode solution brought about a recovery of the membrane potential.

It was interesting that noradrenaline ($2 \cdot 5 \times 10^{-7}$ g. per ml.) caused a marked hyperpolarization when added to the right atrium during late recovery when the resting potential of the cell was declining (*Fig.* 82). Tamai and Kagiyama suggest that this hyperpolarization caused by noradrenaline could be dependent on the electrogenic Na pump, as the effect was too rapid to be explained by passive K redistribution. A similar suggestion for the action of adrenaline was previously proposed by Burnstock (1958). However, it is possible that a membrane resistance change could have been involved.

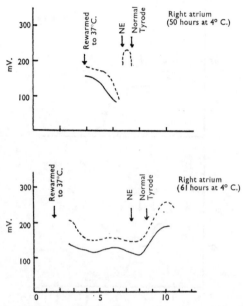

Fig. 82.—Effect of noradrenaline ($2 \cdot 5 \times 10^{-7}$ g. per ml.) on the right atrium. If the noradrenaline was added during rewarming when the resting potential of the cell was in the declining phase, the potential immediately increased. - - -, Maximum; ——, mean.

THE ELECTROGENIC SODIUM PUMP IN TOAD SARTORIUS

MEASURED AND CALCULATED MEMBRANE POTENTIALS

Frumento (1965) obtained results similar to those of Page and Storm (1965). Pairs of sartorius muscles of the South American toad *Leptodactylus ocelatus* were soaked overnight in K-free Ringer at 3° C. to Na-load the fibres. Then both muscles were transferred to normal Ringer solution at 3° C., where values of the membrane potential were recorded. One of the pair of muscles was analysed while the other was warmed to 25° C., and the membrane potential was recorded from a series of fibres for 1 hour. This muscle was then also analysed for Na. As the temperature was raised the membrane potential increased. The change in potential in response to a temperature rise was always greater than that described by the constant-field equation. *Table XIV* shows

Table XIV.—CHANGES IN MEMBRANE POTENTIAL AND EXTRUSION OF SODIUM

EXPERIMENT	EXTRUSION OF Na (mmoles per litre per minute)	MEMBRANE POTENTIAL (mV.)		CORRECTED POTENTIAL (mV.)
		At 3° C.	At 25° C.	
A	0·18	79	96	10
B	0·50	46	92	42
C	0·36	46	76	26
D	0·29	63	86	17
E	0·23	61	76	9
F	0·24	63	77	8
G	0·15	61	73	6
H	0·11	63	77	8

the changes in internal Na concentration per unit time and the membrane potential at 3° C. and 25° C. during a number of experiments. The last column gives maximum corrected potential with temperature taken into account (from the constant-field equation) (Frumento, 1965).

SODIUM EXTRUSION AND POTENTIAL CHANGE

In *Fig.* 83 the corrected changes in potential are plotted against Na extrusion per unit time. This shows that a direct relationship exists between the rate of Na extrusion and the potential change. This would be expected if the potential change were generated by electrogenic Na extrusion.

Fig. 84 shows the membrane potential during recovery, plotted against time. Frumento suggested that the maximum change in potential occurs in so short a time that the internal ion concentrations could not have changed. For example, in B and C of *Fig.* 84 the change in internal K concentration in 6 minutes could be 1·5 mM per litre at the most, which would not explain the change in membrane potential of about 8 mV. which was

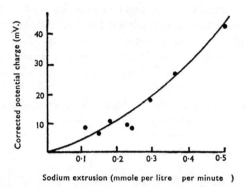

Fig. 83.—Changes in potential plotted against Na extrusion from Na-loaded toad sartorius muscles. There is a linear relationship between corrected potential change and Na extrusion rate. (*Figs.* 83, 84 *reproduced from Frumento, A. S.* (1965), '*Science, N.Y.*', **147**, 1442–1443.)

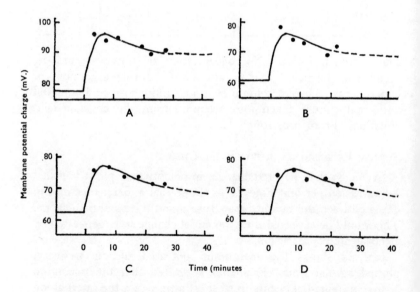

Fig. 84.—Effect of warming on the membrane potential of Na-loaded toad sartorius muscles. (The temperature was changed at zero time.) A rapid increase in membrane potential occurred on warming which was complete after about 6–10 minutes. **A, B, C, D** refer to *Table XIV*, p. 125.

observed in that period. Frumento concluded from these results that the rise in membrane potential during recovery is due to the activity of an electrogenic Na pump as, like Page and Storm (1965), he decided that the rate of change of potential is too great to be due to changes in internal ion concentration during recovery.

Hashimoto (1964) also showed a direct relationship between the rate of Na extrusion and the potential change during recovery of Na-loaded toad sartorius muscles. The membrane potential of the Na-loaded toad muscles was measured at various recovery periods as shown in *Fig.* 85. After each measurement of the

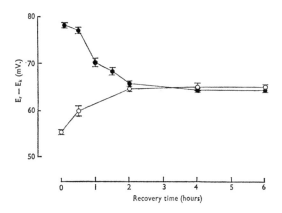

Fig. 85.—Resting potentials (●) and K equilibrium potentials (○) of Na-loaded toad sartorius muscles during recovery. The resting potentials exceeded the E_K levels for nearly 2 hours. (*Figs.* 85-87 *reproduced from Hashimoto, Y.* (1964), '*Kumamoto med. J.*', **18**, 23-30.)

membrane potential, the muscle Na and K were estimated. As can be seen from *Fig.* 85, the membrane potential exceeded the E_K for over 1 hour during recovery. A plot of internal Na concentration to the difference between membrane potential and E_K showed a direct relationship between these two parameters (*Fig.* 86). Since Na efflux is proportional to internal Na concentration, then the rate of Na extrusion is proportional to the potential generated during recovery. This potential difference was reduced by 1 mM ouabain and 2 mM iodoacetate; 0·01 mM 2,4-dinitrophenol had little effect. The effect of these inhibitors is shown in *Fig.* 87.

Hashimoto concluded that since the difference between E_K and the membrane potential increased with increasing Na efflux

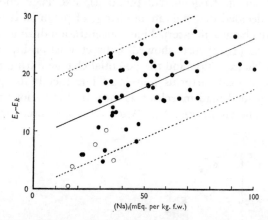

Fig. 86.—Relation between E_r–E_K and the intracellular Na concentration. ○, Fresh muscles. ●, Na-loaded muscles. There is a linear relationship between E_r–E_K and internal Na.

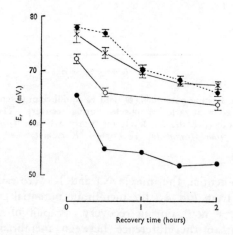

Fig. 87.—Effect of 0·01 mM DNP (×), 2 mM iodoacetate (○), or 1 mM ouabain (●) on the recovery of Na-loaded toad sartorius muscles. (Dotted line represents the control.) Iodoacetate and ouabain reduced the membrane potential during recovery.

part of the membrane potential is created by Na efflux or K influx during recovery, and that the coupling of Na efflux and K influx is not electrically neutral.

The Electrogenic Sodium Pump in Smooth Muscle

An electrogenic Na pump has recently been demonstrated in smooth muscle (Taylor and others, 1969). In fresh pregnant rat myometrium, the mean resting potential was -52 mV. After 18 hours in K-free Krebs bicarbonate solution at $4°$ C., the membrane potential of the Na-loaded muscle-cells fell to -15 mV. Continued immersion after warming to $37°$ C. in K-free Ringer did not increase the membrane potential. However, after addition of 4·6 mM K to the Ringer the membrane potential increased to -70 mV. within the first 2 minutes of exposure to this medium, and exceeded E_K by 5–14 mV. during the early recovery period.

Addition of 10^{-3} M ouabain to the K-free Krebs solution for 10 minutes after Na-loading completely abolished the hyperpolarization normally seen during recovery in K-containing medium at $37°$ C. Cooling the medium from 37 to $25°$ C. retarded the rate of hyperpolarization during recovery.

Taylor and others (1969) suggested that the discrepancy between E_K and the membrane potential during recovery could be due to an electrogenic extrusion of Na producing the hyperpolarization of the smooth muscle-cells.

PART V

THE SIGNIFICANCE OF THE ELECTROGENIC SODIUM PUMP

THE physiological importance of the electrogenic Na pump is greater than was first suspected. This pump mechanism is involved both in the response of neurons to transmitter compounds and also in the production of generator potentials. Other roles have also been suggested for this pump.

ADRENALINE ON FROG SYMPATHETIC GANGLION

The classic response of neurons to transmitter compounds has been described by Eccles (1963). This response is brought about by an increase in membrane permeability to one or more ions. The permeability increase causes the membrane potential to tend towards the resultant equilibrium potential for the ions involved. A change of the membrane potential beyond the equilibrium potential reverses the response. The response can also be reversed by changes of ion levels inside or outside the cell which lead to a change in the equilibrium potential. Recently it has been demonstrated that the response of certain neurons to transmitter compounds is not reversed by potential or ionic changes as described above. Instead the response possesses the properties of an electrogenic Na pump.

MEMBRANE POTENTIAL AND P-POTENTIAL

When orthodromic transmission of frog sympathetic ganglion is blocked by the action of curare or nicotine, a train of repetitive preganglionic stimuli produces a series of potentials (Nishi and Koketsu, 1967a). A short early EPSP is followed by a long-lasting hyperpolarization, the P-potential, and finally by a long-lasting EPSP. The P-potential is attributed to adrenaline.

Nishi and Koketsu (1967a, b) found that this potential possessed unusual properties. Using a sucrose-gap method, they analysed the effects of conditioning depolarization and hyperpolarization on the P-potential. *Fig.* 88 shows the amplitude of the P-potential with regard to the cell membrane potential. A shift of membrane potential of only 10 mV. in either the depolarizing or hyperpolarizing direction caused a large change in amplitude of the P-potential. As shown in *Fig.* 88, a 10-mV. depolarization caused a decrease of about 90 per cent in the size of the P-potential. A 10-mV. hyperpolarization caused an increase

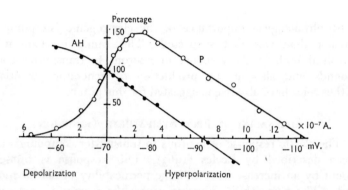

Fig. 88.—Relationship between the P-potential amplitude and the estimated membrane potential of bullfrog sympathetic ganglia. ○, P-potentials. ●, After-hyperpolarizations. Slight hyperpolarization enhances the P-potential. Further hyperpolarization abolishes it. (*Figs.* 88–93 *reproduced from Nishi, S., and Koketsu, K.* (1967b), '*J. Neurophysiol.*', **31**, 717–728.)

of about 50 per cent in the amplitude of the potential. The potential disappeared when the membrane potential was depolarized up to -40 to 60 mV. When the conditioning hyperpolarization was greater than 10–15 mV., the P-potential fell in amplitude and was abolished at membrane potential values ranging from -110 to -135 mV.

These results suggest that the P-potential cannot be generated by movement of ions down their electrochemical gradients. If the potential was due to passive ion movement a depolarization of the membrane potential would be expected to *increase* the amplitude of the potential. Also the very negative value at which the potential disappeared and the lack of reversal make this mechanism unlikely.

RELATION OF THE P-POTENTIAL TO E_K AND E_{Cl}

Further experiments were carried out to show that the P-potential was independent of the E_K of the cell. The E_K was lowered by raising the external K concentration of 10 mM,

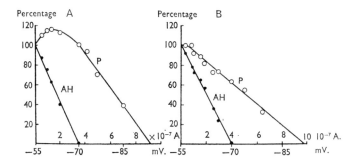

Fig. 89.—Relationship between the P-potential amplitude and the estimated membrane potential of bullfrog sympathetic ganglia. ○, P-potentials. ●, After-hyperpolarizations. The abolition of the P-potential by hyperpolarization occurs at a membrane potential 20 mV. greater than E_K (reversal potential of the after-hyperpolarization.) A and B are records taken from two different preparations.

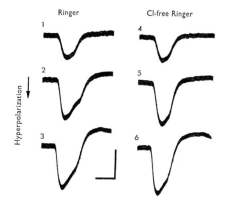

Fig. 90.—Effect of Cl-free Ringer on the P-potentials of a nicotinized ganglion in the presence and absence of a hyperpolarizing current. The P-potential is unchanged by Cl-free Ringer. 2 mV., 5 seconds.

to see the effect on the abolition of P-potential by hyperpolarization (*Fig.* 89). It can be seen that the abolition of the P-potential by a conditioning current occurs at a membrane potential above 20 mV. more negative than the reversal potential of the after-hyperpolarization (E_K).

If the P-potential had been dependent on E_K, a reduction in external K would be expected to *increase* the amplitude. However, a decrease in external K to 0·2 mM *decreased* the amplitude to 60 per cent of that in normal K Ringer (2·0 mM).

The P-potential was also shown to be independent of the E_{Cl} of the membrane (*Fig.* 90). The P-potential showed no change in polarity, size, or duration when the NaCl of the Ringer was totally replaced with equimolar Na glutamate. Also the response of the P-potential to anodal (hyperpolarizing) currents was unchanged.

It was concluded from these results that since the potential was not related to passive ion movements it could be generated by active ion movements. This idea was supported by a number of observations.

High Sodium Perfusion

The preparation was perfused for 35 minutes with Ringer containing twice the normal Na concentration (226·5 mM). The

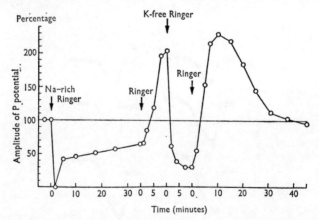

Fig. 91.—Changes in the P-potential amplitude during and after perfusion with Na-rich solution. Na-rich solution leads to a 200 per cent increase in the P-potential which was removed by K-free Ringer.

P-potential disappeared at first but then later reappeared, as shown in *Fig.* 91. When perfusion with normal Ringer was resumed the P-potential immediately increased in amplitude and duration, its amplitude reaching about 200 per cent of the control and its duration increasing 150 per cent over the original. Application of K-free Ringer promptly reduced the P-potential to 30 per cent of the control. By changing back again to normal Ringer the amplitude of the P-potential rose to 230 per cent of the control and then gradually declined.

OUABAIN AND DINITROPHENOL

The association of the P-potential with Na extrusion was supported by earlier work (Nishi and Koketsu, 1967b). *Fig.* 92

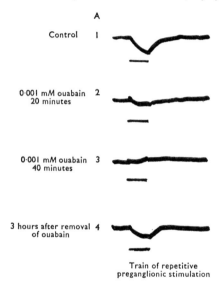

Fig. 92.—Effect of ouabain on the P-potential of a bullfrog sympathetic ganglion cell. Ouabain removed the P-potential after about 20 minutes. Prolonged washing restored the potential.

shows the effect of 10^{-6} M ouabain on the P-potential (*Fig.* 92 A) and after-discharge of the ganglion before application of ouabain. *Fig.* 92 A2, 3 shows the P-potential 20 and 40 minutes after application of ouabain respectively. The P-potential was selectively depressed (*Fig.* 92 A2) and finally abolished (*Fig.* 92 A3). The slow and fast EPSP's were not affected. After 3 hours of washing the effect of ouabain was partially removed.

2,4-Dinitrophenol (10^{-5} or 10^{-4} M) or Na cyanide (10^{-3} M) also reduced the P-potential, but the effect was not as specific as ouabain, since the slow and fast EPSP's were also partly depressed by these inhibitors.

TEMPERATURE AND THE P-POTENTIAL

Changing the temperature has a large effect on the amplitude of the P-potential as shown in *Fig.* 93. By lowering the temperature of the bathing Ringer from 23° C. to 16° C. the P-potential was greatly reduced in amplitude (*Fig.* 93 A). Anodal currents increased the P-potential (*Fig.* 93 B, C) but the size of the potential was still smaller than that at room temperature (*Fig.* 93 D).

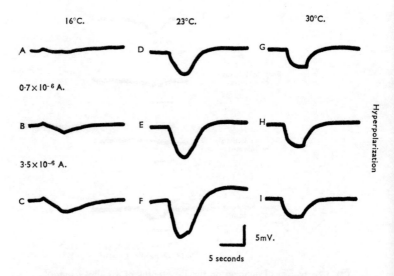

Fig. 93.—Effect of three different temperatures on the P-potentials of a bullfrog sympathetic ganglion cell in the presence and absence of a hyperpolarizing current. The P-potential is reduced by low (16° C.) and high (30° C.) temperature.

By increasing the temperature from 23° C. to 30° C. the P-potential was slightly decreased, and was no longer increased by conditioning hyperpolarization, suggesting that some membrane damage occurs at 30° C.

All these results support the suggestion by Nishi and Koketsu that the P-potential may be produced by the synaptic activation of an electrogenic Na pump in the ganglion cells.

Intracellular Recording and the P-potential

Kobayashi and Libet (1968) also studied the P-potential in frog and rabbit ganglia, but they used direct intracellular recording techniques which gave different results to those obtained by Nishi and Koketsu. No resistance change could be detected during either the slow IPSP (P-potential) or the slow EPSP of the rabbit ganglion cell, in contrast to the fast EPSP's. It is unlikely that this was because there is a brief initial change in resistance giving rise to passive ion movements across the cell membrane, as the time constant of passive decay of the membrane potential is only 10–15 msec. and could therefore not be responsible for such long potentials. The lack of resistance change could not be explained as a generation of the slow potentials at sites remote from the recording electrodes, compared with the fast EPSP, since sympathetic ganglion cells of frog are unipolar and lack dendritic branches.

In curarized rabbit ganglion cells both IPSP and slow EPSP were gradually decreased in amplitude by depolarization and were increased by moderate hyperpolarization. Unlike Nishi and Koketsu, Kobayashi and Libet found that the response of the P-potential to ouabain was *not* specific. In frog and rabbit ganglia the fast EPSP was depressed at about the same slow rate as the IPSP by 10^{-5} M ouabain. It was suggested that this response was probably due to the secondary rise in internal Na and fall in internal K, rather than to a direct effect on the process giving rise to the P-potential.

Kobayashi and Libet were unable to depress the slow IPSP with K-free Ringer. Also the metabolic inhibitors, 2,4-dinitrophenol and azide, and anoxia (95 per cent N_2/5 per cent O_2) had no selective depressant effect on the slow IPSP. In fact the slow EPSP was more sensitive to respiratory inhibitors than the IPSP.

It was concluded that neither the slow EPSP nor the IPSP is generated by ion movements down their electrochemical gradients. However, Kobayashi and Libet were unable to make any suggestion as to how these potentials arose.

In a further study of the hyperpolarizing action of noradrenaline on the mammalian sympathetic ganglion, Kobayashi and Libet (1970) found that the hyperpolarization due to noradrenaline and the depolarization due to the muscarinic action of acetylcholine occurred in the absence of any detectable increase in the ionic conductance of the membrane. This appeared to represent the actual response of the post-synaptic membrane where the drugs acted. The electrogenic action mechanisms for both this noradrenaline response and the acetylcholine response appear to differ from the classic fast synaptic responses in that these slow responses are not due to the movement of ions down their electrochemical gradients.

DOPAMINE ON *Aplysia* NEURONS

Ascher (1968) studied the hyperpolarization by dopamine of recognizable DILDA cells of the abdominal ganglion of *Aplysia*. In the large Br neuron the reversal potential for dopamine hyperpolarization was far more negative than in the adjacent DILDA cells. The difference disappeared after the addition of ouabain. Ascher suggested that part of the hyperpolarizing response to dopamine was due to the stimulation of an electrogenic Na pump.

ACh AND DOPAMINE ON SNAIL NEURONS

Kerkut and others (1969a) reported that the hyperpolarization caused by the addition of dopamine and ACh to snail neurons was inhibited by 10^{-5} g. per ml. ouabain. *Fig.* 94 shows the addition of dopamine to a specified snail neuron. The marked hyperpolarization caused by the addition of dopamine was greatly reduced by ouabain, but the effect was reversible if the ouabain was rapidly washed off. *Fig.* 95 shows a similar experiment on a cell which was hyperpolarized by ACh. Again ouabain reduced the response.

Further work on the ACh response has supported the hypothesis that hyperpolarization could be due to a Na pump mechanism rather than to a passive ion movement following a conductance increase. Kerkut and others (1969b) described the properties of the hyperpolarization produced by ACh in two different recognizable snail neurons. Clear differences were shown between the two types of cholinergic hyperpolarization observed in these cells. Type I showed classic responses (*see* p. 131) while Type II showed properties characteristic of an electrogenic Na pump.

Fig. 96 shows the effect of injecting Cl on the ACh response of the Type I and Type II cells. The Type I response appears to be due to an increase in P_{Cl} since a reduction of the E_{Cl} by Cl injection reverses the IPSP. Cl is not involved in the Type II response.

THE SIGNIFICANCE OF THE SODIUM PUMP

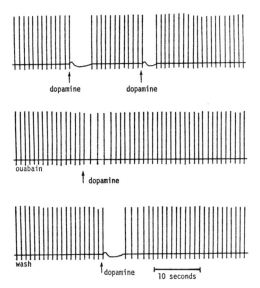

Fig. 94.—The action of dopamine on an electrogenic neuron. Dopamine was added as shown. It caused a hyperpolarization of the nerve-cell and an inhibition of the action potentials. Treatment with 10^{-5} g. per ml. ouabain reduced the effect of dopamine. Washing off the ouabain restored the response to dopamine. Time in seconds. (*Figs.* 94–96 *reproduced from* Kerkut, G. A., Brown, L. C., *and* Walker, R. J. (1969a), '*Life Science*', **8**, 297–300.)

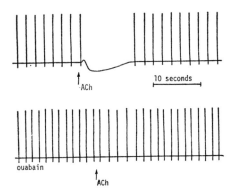

Fig. 95.—The action of acetylcholine (ACh) on an electrogenic neuron. ACh (10^{-5} g. per ml.) hyperpolarized the neuron and the effect was inhibited by ouabain.

Application of ouabain (10^{-4} g. per ml.) had no effect on the Type I response to ACh. However, it reduced and then abolished

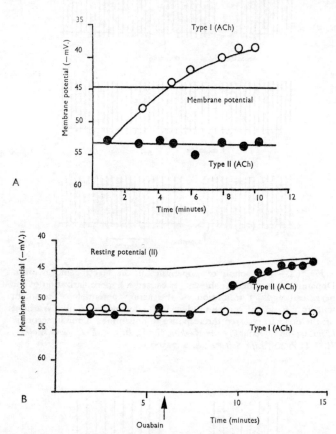

Fig. 96.—The action of acetylcholine (ACh) on a classic neuron and an electrogenic neuron. The effect of injecting chloride ions from a low-resistance KCl-filled electrode on the response to ACh. The classic (Type I) neuron had its response to ACh changed from a hyperpolarization to a depolarization. The Type II neuron (electrogenic) was unaffected by injection of chloride ions. B, The effect of ouabain on the response of the cells to ACh. The Type I cell's response to ACh is unaffected, whilst that of the Type II cell is reduced and abolished.

the Type II hyperpolarization (*Fig.* 97). This suggests that active Na transport is involved in the IPSP generation.

Replacement of external Na by sucrose or Tris–HCl removed the Type II response but not the Type I response.

The Type I response to ACh is the classic type described by Eccles (1963) and is explicable in terms of an increase in P_{Cl}. However, the Type II response is not reversed by injection of Cl, is dependent on the presence of external Na, and is abolished by ouabain. This long-lasting hyperpolarization was reduced by

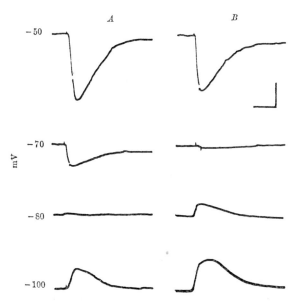

Fig. 97.—The effect of ouabain (2×10^{-4} M) on the K^+-dependent dTC-resistant component of the synaptic inhibition of the pleural neurons in *Aplysia*. A, Shows the post-synaptic response to an intracellular stimulation of the presynaptic neuron in a preparation where dTC (+tubocurarine chloride 10^{-3} g. per ml.) has eliminated the chloride-dependent component. Polarization of the post-synaptic neuron to the levels indicated on the left shows that normally the inversion potential of the dTC-resistant component is -80 mV. B, 10 minutes after the addition of ouabain, the inversion potential was shifted from -80 mV. to -70 mV. Calibration signals: 5 mV., 4 seconds. If the original records had been made at a membrane potential of -70 mV., then the original hyperpolarization would have *apparently* disappeared following application of ouabain. (*Reproduced from Kehoe, J. S., and Ascher, P.* (1970), 'Nature, Lond.', **225**, 820–823.)

tetrodotoxin (10^{-5} g. per ml.) although the action potentials of the cell showing the response were not affected. (Tetrodotoxin affects the passive permeability channels to Na in the membrane.) This suggests that ACh addition to the Type II cell stimulated an electrogenic Na pump by causing an increase in P_{Na} and thus an increase in the intracellular Na concentration. (Since tetrodotoxin has no effect on cell firing here, it seems that Na is possibly not carrying the inward current during the action potentials of this cell.) We do not think that the effect of ouabain on the ACh-induced hyperpolarization in this cell can be explained in terms of the increased K_o produced by stopping the Na–K pump. It is not possible to determine a reversal potential for the ACh-induced hyperpolarization over values of membrane potentials between -35 mV. and -130 mV. The normal ACh-induced hyperpolarization is also relatively unaffected by changes in the external K_o to four times the standard level (4–16 mM).

More recent investigation on the electrogenic hyperpolarization caused by dopamine and noradrenaline (NA) on snail neurons have shown that:—

1. Iontophoretic application of dopamine and NA on to a specified neuron brings about a hyperpolarization of up to 20 mV.
2. Ouabain reduces this hyperpolarization by 50 per cent.
3. The other 50 per cent of the hyperpolarization is probably due to an increase in the membrane permeability to potassium ions.
4. The effect of dopamine or NA is blocked by ergometrine.
5. The cell shows IPSP's which are also blocked by ergometrine and reduced by ouabain.
6. Removal of external sodium ions reduces the dopamine- or NA-induced hyperpolarization.

The system is thus a hybrid one in that the hyperpolarization appears to be partly due to an electrogenic potential and partly to a potassium potential. It is interesting that reducing the external potassium ion concentration increases the hyperpolarization whilst the hyperpolarization is reduced in 10 mM K_o. The potential is sensitive to temperature changes. A decrease in temperature of 17° C. brings a change in the induced hyperpolarization from 18 to 4 mV.

ACh on *Aplysia* Neurons

Pinsker and Kandel (1969) suggested that synaptic activation of an electrogenic Na pump could occur in an identified molluscan interneuron. This neuron is known to cause cholinergic responses

THE SIGNIFICANCE OF THE SODIUM PUMP

in its follower cells by increasing the ionic conductance of the membrane. However, the neuron was found to cause synaptic actions not involving conductance changes. A single action potential in the interneuron was found to give rise to an 800-msec. IPSP in 6 identified follower cells. When the interneuron discharged a train of spikes the duration of the hyperpolarization became greatly prolonged.

Membrane Potential and the Late IPSP

Pinsker and Kandel demonstrated with a number of different experiments that the prolongation of the IPSP involved a second, independent, synaptic process. They termed this process the 'late IPSP' to distinguish it from the 'early IPSP'. First, the early IPSP decreased and finally reversed on hyperpolarization of the membrane potential of the follower cell beyond E_{Cl}. This would be expected if the early IPSP were due to a selective increase in P_{Cl}. However, the late IPSP failed to reverse. Also the early, but not the late, IPSP was blocked by curare. Iontophoretic application of ACh on a follower cell or addition of ACh to the perfusing solution simulated both actions of the interneuron.

Relation of the Late IPSP to E_K and E_{Cl}

The late IPSP was unaffected by Cl-free solution while the early IPSP was inverted. The late IPSP was not increased by a reduction of external K concentration. These results, together with the fact that the IPSP does not reverse on hyperpolarization, suggest that this potential is not due to an increase in conductance to K or Cl. The properties of the late IPSP were consistent with the view that this potential is generated by an electrogenic Na pump. The IPSP was removed by cooling down to 10–7° C. While the IPSP had a magnitude of 3–4 mV. at 19° C., it was nearly absent at 14·2° C. and had disappeared at 10° C. It was also removed by $2 \times 10^{-4} M$ ouabain within 8 minutes. Prolonged washing in K-free solution blocked the late IPSP selectively.

The follower cells showing late IPSP's appeared to possess an electrogenic component of their resting membrane potential. Cooling, ouabain application, and K-free Ringer depolarized these cells by about 10 mV. Also iontophoretic injection of Na hyperpolarized the membrane potential by 10–20 mV., while a comparable injection of K had a very slight effect. This suggests that cells showing late IPSP's might possess an electrogenic component of their membrane potentials with properties similar to those possessed by the IPSP's. It was, however, noted that the late IPSP appeared sensitive to many inhibitors that did *not* act on the sodium pump and so at present one cannot conclude the precise nature of the mechanism of the late IPSP.

As has been described on pp. 50–51, Kehoe and Ascher (1970) present a well-reasoned case for the early and late IPSP to be considered in terms of change in membrane permeability and not in terms of an electrogenic sodium pump. *Fig.* 97 showed the results of one of their experiments concerning the results obtained by Pinsker and Kandel (1969), in which the Na–K pump was stopped by ouabain, and the late IPSP *apparently* disappeared.

ACh AND GABA ON CORTICAL CELLS OF CAT

Krnjevic and Schwartz (1967b) studied the effect of GABA and ACh on some 'unresponsive' cells in the pericruciate cortex of cats. These cells showed a total absence of spontaneous or evoked spikes and synaptic responses. GABA depolarized the cells significantly without causing any change in membrane resistance. ACh also depolarized these cells and some cortical neurons without a fall in membrane resistance. It was suggested that the unresponsive cells were neuroglia and that the depolarizing action of GABA and ACh on the unresponsive cells and neurons indicated a change in electrogenic active transport of Na or Cl.

Godfraind, Krnjevic, and Pumain (1970) found that the action of dinitrophenol on cat cortical neurons was to increase the membrane permeability to potassium and hence inhibit the neuron. The effect of dinitrophenol was not on the metabolism of the nerve-cells but was instead a direct effect on the permeability of the nerve-membrane.

INHIBITORY POTENTIAL OF *Anisodoris*

It is possible that the unusual inhibitory potentials found by Gorman, Mirolli, and Salmoiraghi (1967) in a molluscan neuron are due to electrogenic pump activity. A giant cell in the gastro-oesophageal ganglion of *Anisodoris nobilis* shows an inhibitory potential when the gastro-oesophageal nerve containing its axon is stimulated. The inhibitory potential showed two unusual properties. First, it attained a very large amplitude of 25 mV. with a single stimulus, and a total of 60 mV. during stimulation of the axon at 200 per second. Secondly, it never reversed on hyperpolarization of the soma membrane and, even when the cell was hyperpolarized 60 mV. below the resting potential, the amplitude of the inhibitory potential was unchanged.

MECHANISM OF ACTION OF TRANSMITTER COMPOUNDS

The way in which transmitter compounds stimulate the electrogenic Na pump is not known, although various hypotheses have been put forward. Kerkut and others (1969a, b) have evidence to suggest that the transmitter increases the P_{Na} and causes a hyperpolarization by increasing the internal Na and hence stimulating the pump. Slight falls in membrane resistance have been recorded during the response of the snail neurons to ACh and the response is abolished by reducing the external Na. However, Kobayashi and Libet (1968) and Pinsker and Kandel (1969) noted the lack of a resistance change during the prolonged potentials which they studied. Other suggestions are that the transmitter could stimulate the pump directly (Kerkut and others, 1969a, b) or increase the number of pumping sites, or change the Na-K coupling ratio.

Yet another suggestion is that transmitters could control the electrogenic pump by controlling the energy available to it. Axelsson, Bueding, and Bulbring (1959) and Lundholm and Mohme-Lundholm (1960) suggested that adrenaline causes membrane hyperpolarization by activating cell phosphorylase and thereby increasing the energy available to an electrogenic Na pump.

The advantage of the electrogenic Na pump over a selective increase in ion permeability could be that it can give rise to a long-lasting response, and that this response can be modified by the metabolic state of the cell.

It is possible that long-lasting EPSP's could be generated by an inhibition of the electrogenic Na pump, although no such process has been studied so far.

In a study of the membrane properties of the photoreceptor of the barnacle, Brown, Hagiwara, Koike, and Meech (1970) found that ouabain at a concentration of 10^{-5} M abolished the afterhyperpolarization that follows cessation of illumination. This after-hyperpolarization appears to be due to the activation of an electrogenic sodium pump. The same concentration of ouabain has a negligible effect on the depolarizing receptor potential. There is no evidence that the electrogenic sodium pump is at all involved in the mechanism for the production of the depolarizing receptor potential.

INSECT SENSORY EPITHELIAL POTENTIAL

Thurm (1971) has investigated the potential generated across the sensory epithelium of the campaniform sensilla of the haltere of the fly *Musca*.

The receptor cells are part of the epithelium and each sensilla consists of a sensory cell together with accessory cells (*Fig.* 98).

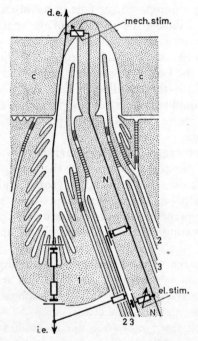

Fig. 98.—Diagram of the sensory receptor cell elements in *Musca*. The main receptor cell (N) is shown centrally with the accessory cells 1, 2, 3 adjacent. The potential develops across A and B. (*After Thurm, U.* (1971), *Verh. dt. zool. Ges.*, **64**, 79–88.)

There appears to be an electrogenic potassium pump which transports potassium ions into the external space. There is no evidence of any low-resistance shunt parallel to the cells of the epithelium. There is a high transepithelial potential of $+50$ to $+100$ mV., the outside being positive. This potential is sensitive

THE SIGNIFICANCE OF THE SODIUM PUMP   147

to anoxia which reduces the potential. There are two main voltage sources suggested for this positive potential. The first is very sensitive to anoxia and is probably in the accessory cell population. The less oxygen-dependent potential is probably on the inner segment of the receptor cell. A mechanical stimulus affects the transepithelial potential via a change in the resistors which are suggested to be in the outer segment of the receptor sensory cell.

The system is similar to the transepithelial potential described by Harvey and his colleagues (Harvey and Nedergaard, 1964; Haskell, Clemons, and Harvey, 1965; Wood, Farrand, and Harvey, 1969) for *Cecropia* midgut. The midgut shows a positive potential of about 110 mV. (the lumen being positive with respect to the blood side of the gut). The potential between the blood side and the inside of a cell is -23 mV. Reduction in Po_2 by passing nitrogen through the bathing solution only affects the positive potential which falls rapidly from $+110$ to about $+14$ mV. There is little or no change in the intracellular potential of -23 mV. (Wood and others, 1969). Isotope studies indicate that the potassium ions carry 83 per cent of the current generated by the midgut when the midgut current is short-circuited (Harvey and Nedergaard, 1964). The short-circuit current is rapidly and reversibly inhibited by anoxia and by 2,4-dinitrophenol. Iodoacetate brought about an irreversible inhibition. Ouabain even at 10^{-4} M did not have any effect on the potential.

It is suggested that sodium ions are not involved in the generation of this potential and that it is probably due to a potassium–hydrogen-linked transport system. It should be noted that in the case of the sensory epithelial potential and the gut potential we are dealing with a potential developed across a cell, whilst most of the nerve and muscle potentials are those developed across a membrane, in other words the inside and the outside of the cell.

STRETCH RECEPTOR NEURON OF CRAYFISH

Nakajima and Onodera (1969) suggested that an electrogenic Na pump could be involved in spike adaptation of the stretch receptor neurons of crayfish. They suggested that the development of spike adaptation could be due to slow changes in membrane permeability brought about by the Na pump, since this could cause a slow increase in the threshold current for spike excitation.

CONTROL OF SYNAPTIC ACTIVITY

Nicholls and Baylor (1968) have suggested another role for the electrogenic Na pump in the control of activity in the brain of leech. Stimulation of the sensory neurons of the leech by electrical or natural (e.g., touch) stimuli causes a hyperpolarization. *Fig.* 99 shows the hyperpolarization in a touch neuron following a train of impulses after the skin had been given a series of touch stimuli.

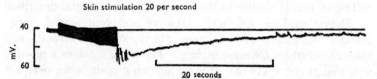

Fig. 99.—Effect of Na injection on the membrane potential of a leech touch neuron. Rapid AP's caused an increase in membrane potential from 42 to 58 mV. (*Figs. 99–103 reproduced from Baylor, D. A., and Nicholls, J. G.* (1969), '*J. Physiol., Lond.*', **203**, 571–589.)

The membrane potential increased from 42 mV. to a peak of 58 mV. after touches were delivered to the skin at 20 per second for 10 seconds by a fine stylus mounted on a piezo-electric element. More intense stimulation caused a larger and slower hyperpolarization of up to 30 mV. which declined over a period of 15–20 minutes.

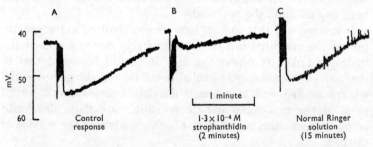

Fig. 100.—Effect of strophanthidin on the hyperpolarization following trains of stimuli applied to the axon of a leech sensory neuron. The amplitude of the hyperpolarization was greatly reduced by the strophanthidin.

The properties of this large hyperpolarization suggested that it was generated by an electrogenic Na pump. It was removed by strophanthidin (4×10^{-4} M). *Fig.* 100 A shows the large control hyperpolarization in response to stimulation. Two minutes after

THE SIGNIFICANCE OF THE SODIUM PUMP

application of the strophanthidin the amplitude of the hyperpolarization was greatly reduced (*Fig.* 100 B). After 15 minutes of washing the response returned (*Fig.* 100 C). No resistance changes were noted during the inhibition by strophanthidin.

The hyperpolarization was also very sensitive to cooling. As no action potentials could be initiated at temperatures below 12° C., the preparation was cooled rapidly as soon as the hyperpolarization began. *Fig.* 101 A shows a control experiment where rapid cooling from 24 to 12° C. depolarized a quiescent neuron

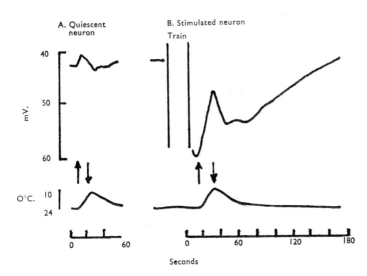

Fig. 101.—Effect of cooling on the hyperpolarization following impulses in a leech sensory neuron. After stimulation, cooling to 12° C. produced a reduction of about 10 mV. in membrane potential.

by about 2 mV. After a train of impulses caused by stimulation of the cell process, cooling to 12° C. now caused a depolarization of about 10 mV. (*Fig.* 101 B).

Baylor and Nicholls discuss two possible implications of this hyperpolarization in activity in leech brain. First, they were able to show that the hyperpolarization recorded in the cell-body extends to the neuropil where synaptic contacts are made. This led to significant changes in integrative activity such as increases in the amplitude of EPSP's, a reversal of IPSP's, and conduction blocks. *Fig.* 102 A shows the response of a quiescent touch cell to stimuli applied to the dorsal root by a suction electrode. This gave rise to three IPSP's which are marked with arrows. If the same stimulus was applied to the root after the cell had been hyperpolarized as a result of stimulation of the skin, the polarity of the IPSP's became reversed (*Fig.* 102 B). The same result was obtained if the cell was hyperpolarized by current injection (*Fig.* 102 C).

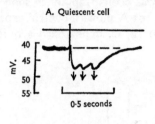

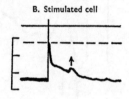

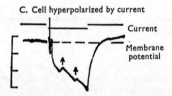

Fig. 102.—Effect of hyperpolarization following stimuli on IPSP's in a leech sensory neuron. Hyperpolarization, either following stimuli or produced by current injection, reversed the IPSP's.

The hyperpolarization also caused conduction block in sensory neurons during and after prolonged stimulation. The spikes which were normally 70 mV. were reduced to about one-third of their amplitude and their time course was prolonged by the hyperpolarization. The configuration was also altered. Nicholls and Baylor suggest that: 'It is therefore likely that the hyperpolarization following activity could similarly block conduction along certain processes, and thereby temporarily "disconnect" the sensory cell from higher order cells with which it normally synapses.'

The hyperpolarization could also control integration in the C.N.S. in a second way. The series of action potentials giving rise to the hyperpolarization would also cause a local rise in external K. This would normally be too small a rise to affect the activity of surrounding cells. However, during the hyperpolarization following a series of action potentials the response of the cell to changes in external K was greatly enhanced. This effect could not be obtained by injection of hyperpolarizing current. Fig. 103 shows the effect of sudden rises in K concentration on a cell which had been hyperpolarized after activity. The glia were removed to minimize delays due to diffusion. Arrows mark the periods during which the external K was increased from 4 to 10 mM. Intracellular recording showed that a short pulse of 10 mM K

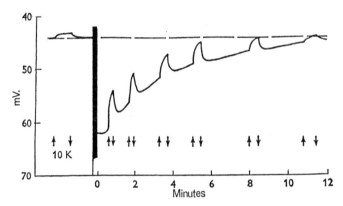

Fig. 103.—Effect of K on the membrane potential of a leech neuron. Sudden increases of external K from 4 to 10 mM depolarized the membrane potential by 8 mV. during hyperpolarization due to electrogenic Na pump activity. These increases decreased the potential by only 2 mV. at the normal resting potential.

Ringer depolarized the cell by only 2 mV. at the normal resting potential. During the hyperpolarization after a train of impulses, a pulse of 10 mM K Ringer caused a 7-mV. depolarization. As there appeared to be no conductance change during K addition, the effect could not be attributed to an increase in K short-circuiting. Baylor and Nicholls suggest instead that the coupling ratio of the Na pump generating the hyperpolarization could be changed by changes in external K (see p. 155).

This rise in K sensitivity during the hyperpolarization could affect integration in the leech C.N.S. as described by Baylor and

Nicholls: 'One remarkable effect of the hyperpolarization produced by activity was to increase the sensitivity of the neuronal membrane potential to K. The possibility therefore exists that neighbouring neurons might interact by such a non-synaptic, K-mediated mechanism. In a situation where the membranes of neuronal processes are closely opposed, one can predict that the consequences of K accumulation will depend on the previous history of the cell. A neuron that has just fired will be more sensitive to K liberated by an adjacent cell than one which is quiescent. According to this scheme activity in one neuronal process could facilitate the recovery of the membrane potential in its neighbours. Such changes in the membrane potential at synapses and regions of low safety factor could influence integration and the conduction of action potentials. Because the changes occur with a relatively slow time course, the interaction would not require precise synchrony of action potentials in the two pathways.'

PIGMENT CONCENTRATION IN PRAWNS

Work carried out by Fingerman (1969) suggests that an electrogenic Na pump may be involved in the control of physiological colour changes in crustaceans. Red chromatophores

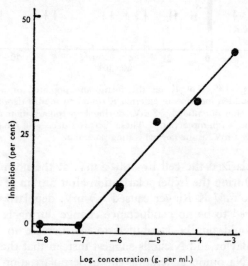

Fig. 104.—Effect of ouabain on the pigment-concentrating response to the pigment-concentrating hormone in the prawn. Increasing the ouabain concentration caused progressive inhibition. (*Reproduced from* Fingerman, M. (1969), '*Am. Zool.*', **9**, 443-452.)

THE SIGNIFICANCE OF THE SODIUM PUMP

(erythrophores) of the prawn *Palaemonetes vulgaris* are concentrated by a pigment-concentrating hormone which depends on the presence of external Na for its action. Ouabain inhibits the response to the hormone. *Fig.* 104 shows that concentrations of ouabain above 10^{-7} g. per ml. ouabain inhibit the response. The hormone hyperpolarizes the membrane potential of the pigment cell, the size of the hyperpolarization being directly related to the degree of pigment concentration. Fingerman suggests that the change in membrane potential and also the change in concentration of the pigment are controlled by a change in the intracellular K/Na ratio, the ratio being changed by stimulation of a neutral Na–K pump by the pigment-concentrating hormone. However, it seems possible that the hyperpolarizations recorded, which are in the order of 13 mV. and are complete within 3 minutes after hormone addition, are so rapid that they are more readily explained in terms of an electrogenic Na pump, which could lead to pigment concentration through the membrane potential change.

PART VI

SOME PROPERTIES OF THE ELECTROGENIC SODIUM PUMP

COUPLING

EXCEPT for the ability of the *electrogenic* Na pump to generate a potential across the membrane of nerve- and muscle-cells, the properties of the pump so far studied are identical to those of the coupled Na–K pump described in these tissues. It is possible that under all conditions Na efflux consists of both electrogenic and electroneutral components, i.e., there is never a 1 : 1 coupling ratio of Na and K movements.

The lack of effect of metabolic inhibitors on the membrane potential of Na-loaded tissues need not exclude the presence of an electrogenic Na pump. This lack of effect could be due to the fact that the membrane resistance of these tissues is too low to allow generation of a measurable potential difference by uncoupled Na efflux. Evidence to support this idea has recently been provided by Shaw and Newby (1970) in giant axons of squid. The addition of metabolic inhibitors to the intact Na-loaded giant axon of squid has no effect on the membrane potential (Hodgkin and Keynes, 1955), i.e., there appears to be no electrogenic component. However, Shaw and Newby found that if the membrane resistance of internally perfused squid axons is increased by perfusion with sucrose-rich medium, the addition of ATP leads to an increase in membrane potential of several millivolts. The external solution was sea water. This preliminary observation suggests that active Na efflux is not entirely coupled in squid axon, and is capable of generating a potential if the membrane resistance is high enough to prevent short-circuiting by the passive movement of ions such as K and Cl.

It has been suggested that Na extrusion could be completely uncoupled to K influx. Kernan (1962a, b) originally suggested that the active excretion of Na could be accompanied only by passive K entrance to restore conditions of electrical neutrality in muscle-cells, during the recovery of Na-loaded muscle. However, the requirement of external K noted later makes this unlikely. In a study on post-tetanic hyperpolarization in myelinated nerve-fibres of frog, Straub (1961) calculated that the hyperpolarization could be explained if only half of the active Na efflux were electrogenic. Further evidence of partial coupling is that K uptake can occur in Na-loaded muscles which exhibit an electrogenic Na pump, when E_K is greater than the membrane potential, i.e., K uptake must be active (Keynes and Rybova, 1963; Cross and others, 1965; Geduldig, 1968a, b).

The stoichiometry of the Na pump has not been widely studied in nerve and muscle, partly because of the difficulty in distinguishing between actively transported K and K which enters down the potential gradient generated by the electrogenic Na pump. However, coupling of Na and K movements by the Na pump of the red blood-cell has been extensively studied. Post and Jolly (1957) estimated that 2 K ions were actively taken up for every 3 Na ions extruded. This ratio has been confirmed by later investigators (Post, Albright, and Dayani, 1967; Garrahan and Glynn, 1967). However, other ratios have also been reported (*see* p. 22).

Nakajima and Takahashi (1966a, b) calculated from the quantity of Na^+ entering crayfish stretch receptor neurons during tetanus, and the electrical quantity responsible for post-tetanic hyperpolarization, that only 20–30 per cent of the total Na^+ extrusion could be coupled to K^+ influx. Thomas (1969) obtained a similar ratio in snail neurons after relating Na^+ efflux to a measurement of the current generated by the pump.

Accurate measurements of the $Na^+ : K^+$ coupling ratio for the Na^+ pump in nerve and muscle are required to discover whether or not this ratio varies under different conditions. The possibility that this might vary has been suggested by Keynes and Rybova (1963). Rang and Ritchie (1968) also suggested that there could be a variable coupling ratio in post-tetanic hyperpolarization in tetanized nerve, since the rate at which the Na^+ debt was discharged varied at different external K^+. However, since the pump current in snail neurons was directly proportional to the rate of Na^+ extrusion, Thomas (1969) suggested that there is a fixed coupling ratio between Na^+ extrusion and presumed K^+ uptake. In the presence of varying external K^+ he found no detectable change in the ratio of pump charge to Na^+ injection charge. However, Rang and Ritchie (1968) suggested that there could be a variable coupling ratio during post-tetanic hyperpolarization in tetanized nerve, since the rate at which the Na debt was discharged varied at different external K concentrations. A variable coupling ratio has also been suggested by Nicholls and Baylor (Nicholls and Baylor, 1968; Baylor and Nicholls, 1969). They noted an increase in sensitivity of the membrane potential to changes in external K during hyperpolarization caused by the electrogenic Na pump in leech sensory neurons. It is possible that when the external K is increased the pump transports K into the cells at a higher rate, i.e., it becomes less electrogenic. Baylor and Nicholls (1969) suggested that the coupling ratio could be controlled by the concentrations of Na and K inside and outside the cell.

The fall in amplitude of post-tetanic hyperpolarization in nerve axons in high external K (10 mM) (Rang and Ritchie, 1968) could also be due to an increase in coupling between Na and K as the external K is raised.

Nakajima and Takahashi (1966b) suggested that the load of the pump might vary the Na–K ratio by a feedback mechanism, a hyperpolarization causing a decrease in the ratio and leading to the enhancement of the pump which they observed.

The idea of a variable coupling ratio has also been put forward to explain the increase in membrane potential by an electrogenic pump in *Aplysia* neurons in response to ACh (Pinsker and Kandel, 1969).

Energy Barriers

The idea of energy barriers was put forward by Conway and others (1961). They reported that there was a critical energy barrier to secretion of Na from frog skeletal muscle of about 2 Kcal. per mole, above which there was no significant secretion of Na and below which Na could be excreted in surprisingly large amounts in a short time. This energy barrier was composed of osmotic and electrical gradients. The energy barrier to Na excretion could be changed by internal and external ion concentrations, by the membrane potential, and by the energy levels of the cell.

The hypothesis that the rate of Na transport is controlled by an energy barrier has been used by a number of workers to explain various properties of the electrogenic Na pump. It was suggested that an energy barrier could explain why active Na efflux increases when external K increases, if there is no coupling of Na to K (Mullins and Awad, 1965). The K concentration would affect the membrane potential, and this could then have a direct effect on the Na pump. However, most evidence points towards coupling of Na and K under all conditions.

Harris and Ochs (1966) found that the membrane potential rise in response to warming Na-loaded muscles was increased by a Na-free medium. They suggested that this was due to the effect of a decreased energy barrier. Kernan (1962a) obtained similar results in Na-loaded muscle and explained them likewise.

Nishi and Koketsu (1967a, b) used a barrier hypothesis to explain the enhancement of the potential generated by the electrogenic Na pump in frog sympathetic ganglia (*see Fig.* 88). They suggested that the effect of membrane on the P-potential was due to a feedback system where more energy was mobilized in the presence of a greater load for the active Na pump. The feedback system could decrease the rate of activity of the Na pump when the load was reduced, which would explain the diminution of electrogenic potential on depolarization. Large hyperpolarizations would exceed a critical barrier and inhibit the pump, as was observed.

Rang and Ritchie (1968) also considered whether there could be a critical energy barrier system for nerve, since the size of the post-tetanic hyperpolarization could not be increased by increasing the duration of stimulation. However, they concluded that this phenomenon was better explained by a saturation of the Na^+ pump at high internal Na^+ concentrations.

A Reconstituted Electrogenic Sodium Pump

In 1957 Skou found an ATPase with unusual properties in the microsomal fraction of crab non-myelinated nerve. It required, besides the presence of Mg^{2+}, the presence of both Na^+ and K^+. Although Na^+ alone, in the presence of Mg^{2+}, has a slight stimulatory effect, while K^+ alone had no effect, the addition of the two ions together caused a large stimulation in the rate of ATP hydrolysis. The stimulation was inhibited by high concentrations of K^+ and also by ouabain. These properties are very similar to those of the mechanism involved in the coupled transport of Na^+ and K^+ in nerve and other membranes. Skou suggested that this enzyme provides the basis for Na^+ and K^+ transport across cell membranes, by energy release on ATP hydrolysis.

SOME PROPERTIES OF THE SODIUM PUMP

Since 1957, similar ATPase preparations from membrane fractions, requiring the presence of Mg^{2+}, Na^+, and K^+, have been found in a wide variety of tissues which carry out Na^+ and K^+ transport, including red blood-cells (Post, Albright, and Dayani, 1960), brain (Hess and Pope, 1957), muscle (Skou, 1962), liver (Emmelot and Bos, 1962), intestine (Taylor, 1962), kidney (Whittam and Wheeler, 1961), and electric organ (Glynn, 1962). A paper by Jain, Strickholm, and Cordes (1969) described the incorporation of a membrane-bound form of a Na–K–ATPase into an artificial membrane. This system was able to generate a current across the membrane.

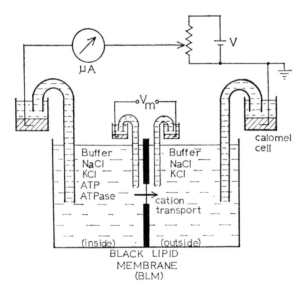

Fig. 105.—Cell for measuring zero voltage membrane current across a black lipid membrane. (*Figs.* 105-7 *reproduced from Jain, M. K., Strickholm, A., and Cordes, E. H.* (1969), '*Nature, Lond.*', **222**, 871–872.)

Jain and others (1969) prepared a black lipid membrane (BLM) support from 1 per cent oxidized cholesterol in octane/duodecane (3 : 2) mixed with approximately 20 parts per million of didodecyl phosphite. This was spread over a 1-mm. diameter hole submerged in a buffer solution containing 0·1 M NaCl, 0·005 M KCl, and 0·005 M $MgCl_2$ with 0·01 M Tris (hydroxymethyl aminomethane), pH 7·35, at 37° C. (*Fig.* 105).

ATP was added to one compartment followed by a preparation of a membrane-bound Na$^+$–K$^+$-dependent ATPase fraction isolated from the synaptic vesicles of rat. This addition led to a

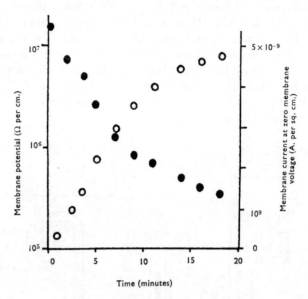

Fig. 106.—The time relation of membrane (slope) resistance (●) and current (○) (at zero membrane voltage) after addition of ATPase to one side of a black lipid membrane.

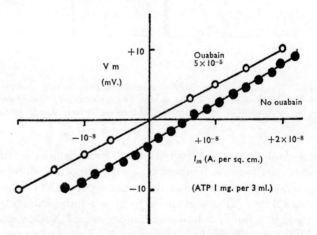

Fig. 107.—The presence of ouabain and related compounds in the medium inhibits the flow of current across a black lipid membrane.

SOME PROPERTIES OF THE SODIUM PUMP

fall in resistance of the BLM and a current flow across the membrane. The current flow was monitored using the short-circuiting method shown in *Fig.* 106, which shows the effect of ATPase addition on the BLM resistance and membrane current at zero voltage. The current flow was interpreted as being due to transport of Na^+ across the membrane, as in the *in vivo* system. This hypothesis was supported by the observation that the current flow was dependent both on the concentration of ATP and on the presence of Na^+ in the medium. Also addition of ouabain and related compounds to the medium inhibited the flow of current as shown in *Fig.* 107.

Jain and others (1969) concluded that if Na^+ and K^+ were the charge-carrying species, it was obvious that the flux of cations in mutually opposite directions was unequal, i.e., the system was not a 1 : 1 Na–K coupled pump. Although the system used was an artificial one, it is tempting to suggest that unequal numbers of Na^+ and K^+ are required on the binding sited for the breakdown of one molecule of ATP and that such a system leads to the voltage-generating properties of the Na^+–K^+ pump described in the previous pages.

PART VII
GENERAL CONCLUSION

THE evidence presented in this account indicates that in some nerve-cells and axons the activity of the sodium pump can contribute to the potential across the nerve membrane. Just how important this electrogenic potential is in the activity and control of the neurons in the C.N.S. is at present hard to tell. The present account has accentuated the positive evidence in favour of the system. Quite clearly the sodium pump will be more important in those axons of small size (C fibres) or those axons where for some reason there is a high rate of sodium entry. Whether a potential is produced by the pump may depend on features such as the membrane resistance. Whether there can be variable coupling between the sodium pumped out and the potassium pumped in is still to be determined.

There is also some evidence that the electrogenic pumps can be affected by chemicals present around the neuron; thus some transmitters and metabolites affect the pump activity and hence the potential. It is likely that this type of activity may prove to be of importance in the control of longer-term reactions in the C.N.S.

It is also probable that other electrogenic pumps could be involved in the movements of ions, such as sodium, potassium, chloride, calcium, lactate, glutamate, hydrogen, etc., and the action of these pumps may affect both the membrane potential and also the physical properties of the nerve membrane. These pumps provide a link between the biochemistry and the electrical activity of the nerve-cell and may also indicate the mechanisms of actions of drugs on the neurons. The drugs could affect the systems providing the energy for the pump or they could act specifically on the pump mechanism itself.

By the study of the action of these pumps one is able to link neurochemistry and electrophysiology and this should lead to a more complete understanding of the complex behaviour of the cells within the central nervous system.

REFERENCES

ADRIAN, R. H., and FREYGANG, W. H. (1962), 'The potassium and chloride conductance of frog muscle membrane', *J. Physiol., Lond.*, **163**, 61–103.

ADRIAN, R. H., and SLAYMAN, C. L. (1964), 'Pumped movements of K and Rb in frog muscle', *J. Physiol., Lond.*, **175**, 49–50P.

ADRIAN, R. H., and SLAYMAN, C. L. (1966), 'Membrane potential and conductance during transport of sodium, potassium, and rubidium in frog muscle', *J. Physiol., Lond.*, **184**, 970–1014.

ALVING, B., and CARPENTER, D. (1967), 'The significance of an electrogenic Na pump in *Aplysia* neurons', *Fedn Proc. Fedn Am. Socs exp. Biol.*, **26**, 240.

ASCHER, P. (1968), 'Electrophoretic injections of dopamine on *Aplysia* neurones', *J. Physiol., Lond.*, **198**, 48–49P.

AXELSSON, J., BUEDING, E., and BULBRING, E. (1959), 'The action of adrenaline on phosphorylase activity and membrane potential of smooth muscle', *J. Physiol., Lond.*, **148**, 62–63P.

BAKER, P. F. (1965), 'Phosphorus metabolism of intact crab nerve and its relation to the active transport of ions', *J. Physiol., Lond.*, **180**, 383–423.

BAKER, P. F. (1966), 'The sodium pump', *Endeavour*, **96**, 166–169.

BAKER, P. F., BLAUSTEIN, M. P., KEYNES, R. D., MANIL, J., SHAW, T. I., and STEINHARDT, R. A. (1969), 'The ouabain-sensitive fluxes of sodium and potassium in squid giant axons', *J. Physiol., Lond.*, **200**, 459–496.

BAYLOR, D. A., and NICHOLLS, J. G. (1969), 'After effects of nerve impulses on signalling in the central nervous system of the leech', *J. Physiol., Lond.*, **203**, 571–589.

BERNSTEIN, J. (1902), 'Untersuchungen zur Thermodynamik der biologischen Strome', *Pflügers Arch. ges. Physiol.*, **92**, 521–537.

BERNSTEIN, J. (1912), *Electrobiologie*. Braunschweig: Vieweg.

BOHM, W. H., and STRAUB, R. W. (1961), 'Effects of cardiac glycosides on the hyperpolarization which follows activity in nerve fibres', in *New Aspects of Cardiac Glycosides, First International Pharmacology Meeting* (ed. WILBRANDT, W.), vol. 3, pp. 245–251. Oxford: Pergamon.

BONTING, S. L. (1970), 'Sodium potassium activated adenosine triphosphatase and cation transport', in *Membranes and Ion Transport* (ed. BITTAR, E. E.), vol. 1, pp. 257–363. New York: Wiley.

BOURKE, R. S., and TOWER, D. B. (1966), 'Fluid compartmentation and electrolytes of cat cerebral cortex *in vitro*. II. Sodium, potassium and chloride of mature cerebral cortex', *J. Neurochem.*, **13**, 1099–1117.

BOYLE, P. F., and CONWAY, E. J. (1941), 'Potassium accumulation in muscle and associated changes', *J. Physiol., Lond.*, **100**, 1–63.

BROWN, H. M., HAGIWARA, S., KOIKE, H., and MEECH, R. M. (1970), 'Membrane properties of a barnacle photoreceptor examined by voltage clamp technique', *J. Physiol., Lond.*, **208**, 385–414.

BURNSTOCK, G. (1958), 'The action of adrenaline on, and excitability and membrane potential in, the taenia coli of the guinea pig and the effect of DNP on this action and on the action of acetylcholine', *J. Physiol., Lond.*, **143**, 183–194.

CALDWELL, P. C., HODGKIN, A. L., KEYNES, R. D., and SHAW, T. I. (1960a), 'The effects of injecting "energy-rich" phosphate compounds on the active transport of ions in the giant axons of *Loligo*', *J. Physiol., Lond.*, **152**, 561–590.

CALDWELL, P. C., HODGKIN, A. L., KEYNES, R. D., and SHAW, T. I. (1960b), 'Partial inhibition of the active transport of cations in the giant axons of *Loligo*', *J. Physiol., Lond.*, **152**, 591–600.

CALDWELL, P. C., and KEYNES, R. D. (1959), 'The effect of ouabain on the efflux of sodium from a squid giant axon', *J. Physiol., Lond.*, **148**, 8–10P.

CALDWELL, P. C., and KEYNES, R. D. (1960), 'The permeability of the squid giant axon to radioactive potassium and chloride ions', *J. Physiol., Lond.*, **154**, 177–189.

CAREY, M. J., CONWAY, E. J., and KERNAN, R. P. (1959), 'Secretion of sodium ions by the frog's sartorious', *J. Physiol., Lond.*, **148**, 51–82.

CARPENTER, D. O. (1967), 'Temperature effects on pacemaker generation, membrane potential, and critical firing threshold in *Aplysia* neurons', *J. gen. Physiol.*, **50**, 1469–1484.

CARPENTER, D. O. (1970), 'Membrane potential produced directly by the Na pump in *Aplysia* neurones', *Comp. Biochem. Physiol.*, **35**, 371–383.

CARPENTER, D. O., and ALVING, B. O. (1968), 'A contribution of an electrogenic Na pump to the membrane potential in *Aplysia* neurones', *J. gen. Physiol.*, **52**, 1–21.

CHALAZONITIS, N. (1961), 'Chemopotentials in giant nerve cells (*Aplysia fasciata*)', in *Nervous Inhibition* (ed. FLOREY, E.), pp. 179–193. London: Pergamon.

CHALAZONITIS, N., GOLA, M., and ARVANITAKI, A. (1965), 'Oscillations lentes du potentiel de membrane neuronique, fonction de la Po_2 intracellulaire (neurones autoactifs d'*Aplysia depilans*)', *C. r. Seanc. Soc. Biol.*, **159**, 2451–2455.

CONNELLY, C. M. (1959), 'Recovery processes and metabolism of nerve', *Rev. mod. Phys.*, **31**, 475–484.

CONWAY, E. J. (1964), 'New light on the active transport of sodium ions from skeletal muscle', *Fedn Proc. Fedn Am. Socs exp. Biol.*, **23**, 680–688.

CONWAY, E. J., KERNAN, R. P., and ZADUNAISKY, J. A. (1961), 'The sodium pump in skeletal muscle in relation to energy barriers', *J. Physiol., Lond.*, **155**, 263–279.

COOMBS, J. S., ECCLES, J. C., and FATT, P. (1955a), 'The electrical properties of the motoneurone membrane', *J. Physiol., Lond.*, **130**, 291–325.

COOMBS, J. S., ECCLES, J. C., and FATT, P. (1955b), 'Excitatory synaptic action in motoneurones', *J. Physiol., Lond.*, **130**, 374–395.

CROSS, S. B., KEYNES, R. D., and RYBOVA, R. (1965), 'The coupling of sodium efflux and potassium influx in frog muscle', *J. Physiol., Lond.*, **181**, 865–880.

DEAN, R. B. (1941), 'Theories of electrolyte equilibrium in muscle', *Biol. Symp.*, **3**, 331–348.

DESMEDT, J. E. (1953), 'Electrical activity and intracellular sodium concentration in frog muscle', *J. Physiol., Lond.*, **121**, 191–205.

REFERENCES

DONNAN, F. G. (1924), 'The theory of membrane equilibria', *Chem. Rev.*, **7**, 73-90.

ECCLES, J. C. (1963), *The Physiology of Synapses*. Berlin: Springer.

ECCLES, R. M., LØYNING, Y., and OSHIMA, T. (1966), 'Effects of hypoxia on the monosynaptic reflex pathway in the cat spinal cord', *J. Neurophysiol.*, **29**, 315-332.

EISENMAN, G., RUDIN, D. O., and CASBY, J. U. (1957), 'Glass electrode for measuring sodium ion', *Science, N.Y.*, **126**, 831-834.

EMMELOT, P., and BOS, C. J. (1962), 'Adenosine triphosphatase in the cell membrane fraction from rat liver', *Biochim. biophys. Acta*, **58**, 374-375.

FENN, W. O. (1962), 'Born fifty years too soon', *Ann. Rev. Physiol.*, **24**, 1-10.

FENN, W. O., and COBB, D. M. (1934), 'The potassium equilibrium in muscle', *J. gen. Physiol.*, **17**, 629-656.

FINGERMAN, M. (1969), 'Cellular aspects of the control of physiological colour changes in crustaceans', *Am. Zool.*, **9**, 443-452.

FLEAR, C. T. G. (1962), 'Muscle biopsy and appraisal of cellular potassium; observation on rats', *Proc. Ass. Clin. Biol.*, **2**, 59-62.

FLEAR, C. T. G., and FLORENCE, I. (1963), 'Muscle biopsy in man; an index of cellular potassium?', *Nature, Lond.*, **199**, 156-158.

FRUMENTO, A. S. (1965), 'Sodium pump: Its electrical effects in skeletal muscle', *Science, N.Y.*, **147**, 1442-1443.

GARRAHAN, P. J., and GLYNN, I. M. (1967), 'The stoichiometry of the sodium pump', *J. Physiol., Lond.*, **192**, 217-235.

GEDULDIG, D. (1968a), 'A ouabain-sensitive membrane conductance', *J. Physiol., Lond.*, **194**, 521-533.

GEDULDIG, D. (1968b), 'Analysis of membrane permeability coefficient ratios and internal ion concentrations from a constant field equation', *J. theor. Biol.*, **19**, 67-78.

GIBBS, WILLARD J. (1906), *The Selected Scientific Papers of J. Willard Gibbs*. London: Longmans.

GLYNN, I. M. (1956), 'Sodium and potassium movements in human red cells', *J. Physiol., Lond.*, **134**, 278-310.

GLYNN, I. M. (1957a), 'The ionic permeability of the red cell membrane', *Prog. Biophys.*, **8**, 242-307.

GLYNN, I. M. (1957b), 'The action of cardiac glycosides on sodium and potassium movements in human red blood cells', *J. Physiol., Lond.*, **136**, 148-173.

GLYNN, I. M. (1962), 'Activation of adenosinetriphosphatase activity in a cell membrane by external potassium and internal sodium', *J. Physiol., Lond.*, **160**, 18P.

GODFRAIND, J. M., KRNJEVIC, K., and PUMAIN, R. (1970), 'Unexpected features of the action of dinitrophenol on cortical neurones', *Nature, Lond.*, **228**, 562-564.

GORMAN, A. L. F., and MARMOR, M. (1970a), 'Contributions of the sodium pump and ionic gradients to the membrane potential of a molluscan neurone', *J. Physiol., Lond.*, **210**, 897-918.

GORMAN, A. L. F., and MARMOR, M. (1970b), 'Temperature dependence of a sodium potassium permeability ratio of a molluscan neurone', *J. Physiol., Lond.*, **210**, 919–932.

GORMAN, A. L. F., MIROLLI, M., and SALMOIRAGHI, G. C. (1967), 'Unusual characteristics of an inhibitory potential recorder intracellularly from a mollusc nerve cell', *Fedn Proc. Fedn Am. Socs exp. Biol.*, **27**, 329.

GRAHAM, J., and GERARD, R. W. (1946), 'Membrane potentials and excitation of impaled single muscle fibres', *J. cell. comp. Physiol.*, **28**, 99–117.

GREENGARD, P., and STRAUB, R. W. (1962), 'Metabolic studies on the hyperpolarisation following activity in mammalian non-myelinated nerve fibres', *J. Physiol., Lond.*, **161**, 414–423.

GRUNDFEST, H. (1955), 'The nature of electrochemical potentials of bioelectric tissues', in *Electrochemistry in Biology and Medicine* (ed. SHELDOVSKY, T.). New York: Wiley.

GRUNDFEST, H., KAO, C. Y., and ALTAMIRANO, M. (1955), 'Bioelectric effects of ions microinjected into the giant axon of *Loligo*', *J. gen. Physiol.*, **38**, 245–282.

GURBER, G. (1895), quoted in GELLHORN, E. (1929), *Das Permeabilitätsproblem*. Berlin: Springer.

HAGIWARA, S., GRUENER, R., HAYASHI, H., SAKATA, H., and GRINNEL, A. D. (1968), 'The effect of external and internal pH changes on K and Cl conductances in a muscle fibre membrane of the giant barnacle', *J. gen. Physiol.*, **52**, 773–792.

HALDANE, J. S., and PRIESTLEY, J. G. (1935), *Respiration*, p. vii. Oxford: Clarendon.

HAMBURGER, H. J. (1891), 'Über den Einfluss der Athmung auf die Permeabilität der Blutkörperchen', *Z. Biol.*, **28**, 405–416.

HARRIS, E. J. (1954), 'Linkage of Na and K active transport in human erythrocytes', *Symp. Soc. exp. Biol.*, **8**, 228–241.

HARRIS, E. J., and MAIZELS, M. (1951), 'The permeability of human erythrocytes to Na', *J. Physiol., Lond.*, **113**, 506–524.

HARRIS, J. E. (1941), 'The influence of the metabolism of human erythrocytes on their potassium content', *J. biol. Chem.*, **141**, 579–595.

HARRIS, J. E., and OCHS, S. (1966), 'Effects of sodium extrusion and local anaesthetics on muscle membrane resistance and potential', *J. Physiol., Lond.*, **187**, 5–21.

HARVEY, W. R., and NEDERGAARD, S. (1964), 'Sodium independent active transport of potassium in the isolated midgut of the *Cecropia* silkworm', *Proc. natn. Acad. Sci. U.S.A.*, **51**, 757–765.

HASHIMOTO, Y. (1964), 'Resting potentials of Na-loaded sartorious muscle fibres of toads during recovery in high K Ringer', *Kumamoto med. J.*, **18**, 23–30.

HASKELL, J. A., CLEMONS, R. D., and HARVEY, W. R. (1965), 'Active transport by the *Cecropia* midgut. Inhibitors, stimulants and potassium transport', *J. cell. comp. Physiol.*, **65**, 45–56.

HEPPEL, L. A. (1939), 'The electrolytes of muscle and liver in potassium depleted rats', *Am. J. Physiol.*, **127**, 385–392.

REFERENCES

HEPPEL, L. A. (1940), 'Diffusion of radioactive sodium into the muscles of potassium deprived rats', *Am. J. Physiol.*, **128**, 449–454.

DEN HERTOG, A., GREENGARD, P., and RITCHIE, J. M. (1969), 'On the metabolic basis of nervous activity', *J. Physiol., Lond.*, **204**, 511–522.

DEN HERTOG, A., and RITCHIE, J. M. (1969), 'A comparison of the effects of temperature, metabolic inhibitors and of ouabain on the electrogenic component of the sodium pump in mammalian non-myelinated nerve fibres', *J. Physiol., Lond.*, **204**, 523–538.

HESS, H. H., and POPE, A. (1957), 'Effect of metal cations on adenosine triphosphatase activity of rat brain', *Fedn Proc. Fedn Am. Socs exp. Biol.*, **16**, 196.

HINKE, J. A. M. (1961), 'The measurement of sodium and potassium activities in the squid axon by means of cation-selective glass micro electrodes', *J. Physiol., Lond.*, **156**, 314–335.

HÖBER, R. (1945), *Physical Chemistry of Cells and Tissues*. New York: Blakiston.

HODGKIN, A. L. (1958), 'Ionic movements and electrical activity in giant nerve fibres', *Proc. R. Soc. B*, **148**, 1–37.

HODGKIN, A. L., and HOROWICZ, P. (1959), 'The influence of potassium and chloride ions on the membrane potential of single muscle fibres', *J. Physiol., Lond.*, **148**, 127–160.

HODGKIN, A. L., and HOROWICZ, P. (1960), 'The effect of sudden changes in ionic concentration on the membrane potential of single muscle fibres', *J. Physiol., Lond.*, **153**, 370–385.

HODGKIN, A. L., and HUXLEY, A. F. (1945), 'Resting and action potentials in single nerve fibres', *J. Physiol., Lond.*, **104**, 176–195.

HODGKIN, A. L., and KEYNES, R. D. (1955), 'Active transport of cations in giant axons from *Sepia* and *Loligo*', *J. Physiol., Lond.*, **128**, 28–60.

HODGKIN, A. L., and KEYNES, R. D. (1956), 'Experiments on injection of substances into squid giant axons by means of a microsyringe', *J. Physiol., Lond.*, **131**, 592–616.

HOFFMAN, J. F., TOTESON, D. C., and WHITTAM, R. (1960), 'Retention of potassium by human erythrocyte ghosts', *Nature, Lond.*, **185**, 186–187.

HOFFMAN, J. F. (1962), 'Cation transport and structure of the cell plasma membrane', *Circulation*, **26**, 1201–1213.

HOKIN, L. (1969), 'On the molecular characterisation of the sodium potassium transport adenosine triphosphatase', *J. gen. Physiol.*, **54**, 327s–342s.

HOLMES, O. (1962), 'Effects of pH, changes in potassium concentration and metabolic inhibitors on the after-potentials of mammalian non-medullated nerve fibres', *Archs int. Physiol. Biochim.*, **70**, 211–245.

ITO, M., and OSHIMA, T. (1965), 'The electrogenic action of cations on cat spinal motoneurones', *Proc. R. Soc. B*, **161**, 92–108.

JAIN, M. K., STRICKHOLM, A., and CORDES, E. H. (1969), 'Reconstitution of an ATP-mediated active transport system across black lipid membranes', *Nature, Lond.*, **222**, 871–872.

KAPITSA, P. C. (1964), 'The future problems of science', in *The Science of Science* (ed. GOLDSMITH, M., and MACKAY, A.). London: Souvenir Press.

KEHOE, J. S. (1967), 'Pharmacological characteristics and ionic bases of a two component post synaptic inhibition', *Nature, Lond.*, **215**, 1503–1505.

KEHOE, J. S. (1969), 'Single presynaptic neurone mediates a two component post synaptic inhibition', *Nature, Lond.*, **221**, 866–868.

KEHOE, J. S., and ASCHER, P. (1970), 'Re-evaluation of the synaptic activation of an electrogenic sodium pump', *Nature, Lond.*, **225**, 820–823.

KERKUT, G. A. (1967), 'Biochemical aspects of invertebrate nerve cells', in *Invertebrate Nervous Systems* (ed. WIERSMA, C. A. G.), pp. 5–37. Chicago University Press.

KERKUT, G. A., BROWN, L. C., and WALKER, R. J. (1969a), 'Post synaptic stimulation of the electrogenic sodium pump', *Life Science*, **8**, 297–300.

KERKUT, G. A., BROWN, L. C., and WALKER, R. J. (1969b), 'Cholinergic stimulation of the electrogenic sodium pump', *Nature, Lond.*, **223**, 864–865.

KERKUT, G. A., and RIDGE, R. M. A. P. (1961), 'The effect of temperature changes on the resting potential of crab, insect and frog muscle', *Comp. Biochem. Physiol.*, **3**, 64–70.

KERKUT, G. A., and RIDGE, R. M. A. P. (1962), 'The effect of temperature changes on the activity of the neurones of the snail *Helix aspersa*', *Comp. Biochem. Physiol.*, **5**, 283–295.

KERKUT, G. A., and THOMAS, R. C. (1963), 'Acetylcholine and the spontaneous inhibitory post synaptic potentials in the snail neurone', *Comp. Biochem. Physiol.*, **8**, 39–45.

KERKUT, G. A., and THOMAS, R. C. (1964), 'The effect of anion injection and changes in the external potassium and chloride concentration on the reversal potentials of the IPSP and acetylcholine', *Comp. Biochem. Physiol.*, **11**, 199–213.

KERKUT, G. A., and THOMAS, R. C. (1965), 'An electrogenic sodium pump in snail nerve cells', *Comp. Biochem. Physiol.*, **14**, 167–183.

KERKUT, G. A., and YORK, B. (1969), 'The oxygen sensitivity of the electrogenic sodium pump in snail neurones', *Comp. Biochem. Physiol.*, **28**, 1125–1134.

KERNAN, R. P. (1962a), 'Membrane potential changes during sodium transport in frog sartorius muscle', *Nature, Lond.*, **193**, 986–987.

KERNAN, R. P. (1962b), 'The role of lactate in the active excretion of sodium by frog muscle', *J. Physiol., Lond.*, **162**, 129–137.

KERNAN, R. P. (1966), 'Denervation and the electrogenesis of the sodium pump in frog skeletal muscle', *Nature, Lond.*, **210**, 537–538.

KERNAN, R. P., and TANGNEY, A. (1964), 'An electrogenic Na-pump in frog striated muscle', *J. Physiol., Lond.*, **172**, 32P.

KEYNES, R. D., and LEWIS, P. R. (1951), 'The sodium and potassium content of cephalopod nerve fibres', *J. Physiol., Lond.*, **114**, 151–182.

KEYNES, R. D., and RYBOVA, R. (1963), 'The coupling between sodium and potassium fluxes in frog sartorius muscle', *J. Physiol., Lond.*, **168**, 58P.

REFERENCES

KEYNES, R. D., and SWAN, R. C. (1959), 'The permeability of frog muscle fibres to lithium ions', *J. Physiol., Lond.*, **147**, 626–638.

KHAN, J. B. (1958), 'Relations between calcium and potassium transfer in human erythrocytes', *J. Pharmac. exp. Ther.*, **123**, 263–268.

KOBAYASHI, H., and LIBET, B. (1968), 'Generation of slow postsynaptic potentials without increases in ionic conductance', *Proc. Natn. Acad. Sci. U.S.A.*, **60**, 1304–1311.

KOBAYASHI, H., and LIBET, N. B. (1970), 'Actions of noradrenaline and acetylcholine on sympathetic ganglion cells', *J. Physiol., Lond.*, **208**, 385–414.

KOECHLIN, B. A. (1955), 'On the chemical composition of the axoplasm of giant nerve fibres with particular reference to its ion pattern', *J. Biophys. Biochem. Cytol.*, **1**, 511–529.

KOLMODIN, G. M., and SKOGLUND, C. R. (1959), 'Influence of asphyxia on membrane potential level and action potentials of spinal moto- and interneurones', *Acta physiol. scand.*, **45**, 1–18.

KOSTYUK, P. G., SOROKINA, Z. A., and KHOLODOVA, YU. D. (1969), 'Measurements of activity of hydrogen, potassium and sodium ions in striated muscle fibres and nerve cells', in *Glass Microelectrodes* (ed. LAVALEE, M., SCHANNE, O., and HEBERT, N. C.), pp. 322–348. New York: Wiley.

KRNJEVIC, K., and SCHWARTZ, S. (1967a), 'Some properties of unresponsive cells in the cerebral cortex', *Exp. Brain Res.*, **3**, 306–319.

KRNJEVIC, K., and SCHWARTZ, S. (1967b), 'The action of GABA on cortical neurones', *Exp. Brain Res.*, **3**, 320–336.

KUNO, M., MIYAHARA, J. T., and WEAKLY, J. N. (1970), 'Post tetanic hyperpolarization produced by an electrogenic pump in dorsal spinocerebellar tract neurones of the cat', *J. Physiol., Lond.*, **210**, 839–855.

LAVALLEE, M., SCHANNE, O., and HEBERT, N. C. (1969), *Glass Microelectrodes*. New York: Wiley.

LEV, A. A. (1964), 'Determination of activity and activity coefficients of potassium and sodium ions in frog muscle fibres', *Nature, Lond.*, **201**, 1132–1134.

LLOYD, P. C. (1953), 'Influence of asphyxia upon the responses of spinal motoneurons', *J. gen. Physiol.*, **36**, 673–702.

LUNDHOLM, L., and MOHME-LUNDHOLM, E. (1960), 'The action of adrenaline on carbohydrate metabolism in relation to some of its pharmacodynamic effects', in *Adrenergic Mechanisms: Ciba Foundation Symposium* (ed. VANE, J. R.). London: Churchill.

MCCONAGHEY, P. D., and MAIZELS, M. (1962), 'Cation exchanges of lactose-treated human red cells', *J. Physiol., Lond.*, **162**, 485–509.

MAIZELS, M. (1954), 'Active transport in erythrocytes', *Symp. Soc. exp. Biol.*, **8**, 202–227.

MAIZELS, M., and PATTERSON, J. H. (1940), 'Survival of stored blood after transfusion', *Lancet*, **2**, 417–420.

MARMOR, M., and GORMAN, A. L. F. (1970), 'Membrane potential as the sum of ionic and metabolic components', *Science, N.Y.*, **167**, 65–67.

MICHAELIS, L. (1928), 'Contributions to the theory of permeability of membranes for electrolytes', *J. gen. Physiol.*, **8**, 33–59.

MORETON, R. B. (1968), 'An application of the constant-field theory to the behaviour of giant neurones of the snail *Helix aspersa*', *J. exp. Biol.*, **48**, 611–623.

MORETON, R. B. (1969), 'An investigation of the electrogenic Na pump in snail neurones, using the constant field theory', *J. exp. Biol.*, **51**, 181–201.

MULLINS, L. J., and AWAD, M. Z. (1965), 'The control of the membrane potential of muscle fibers by the sodium pump', *J. gen. Physiol.*, **48**, 761–775.

MURRAY, R. W. (1966), 'The effect of temperature on the membrane properties of neurons in the visceral ganglion of *Aplysia*', *Comp. Biochem. Physiol.*, **18**, 291–303.

NAKAJIMA, S., and ONODERA, K. (1969), 'Membrane properties of the stretch receptor neurones of crayfish with particular reference to mechanisms of sensory adaptation', *J. Physiol., Lond.*, **200**, 161–185.

NAKAJIMA, S., and TAKAHASHI, K. (1965), 'Post-tetanic hyperpolarisation on stretch receptor neuron of crayfish', in *Symposium on Comparative Neurophysiology* (ed. KATSUKI, Y.). Tokyo: University of Tokyo.

NAKAJIMA, S., and TAKAHASHI, K. (1966a), 'Post-tetanic hyperpolarisation in stretch receptor neurone of crayfish', *Nature, Lond.*, **209**, 1220–1222.

NAKAJIMA, S., and TAKAHASHI, K. (1966b), 'Post-tetanic hyperpolarization and electrogenic Na pump in stretch receptor neurone of crayfish', *J. Physiol., Lond.*, **187**, 105–127.

NETTER, H. (1928), 'Über die Electrolytgleichgewichte an elektiv ionenpermeabelen Membranen und ihre biologische Bedeutung', *Pflügers Arch. ges. Physiol.*, **220**, 107–123.

NICHOLLS, J. G., and BAYLOR, D. A. (1968), 'Long-lasting hyperpolarisation after activity of neurones in leech central nervous system', *Science, N.Y.*, **162**, 279–281.

NISHI, S., and KOKETSU, K. (1967a), 'Excitatory and inhibitory postsynaptic potentials of amphibian sympathetic ganglion cells', *Fedn Proc. Fedn Am. Socs exp. Biol.*, **26**, 329.

NISHI, S., and KOKETSU, K. (1967b), 'Origin of ganglionic inhibitory post synaptic potential of bullfrog sympathetic ganglion', *J. Neurophysiol.*, **31**, 717–728.

NISHI, S., and SOEDA, H. (1964), 'Hyperpolarisation of a neurone membrane by barium', *Nature, Lond.*, **204**, 761–764.

OBARA, S., and GRUNDFEST, H. (1968), 'Effects of lithium on different membrane components of crayfish stretch receptor neurons', *J. gen. Physiol.*, **51**, 635–654.

OVERTON, E. (1895) 'Über die osmotischen Eigenschaften der lebenden Pflanzen und Tierzelle', *Vjschr. naturf. Ges. Zürich*, **40**, 159–201.

PAGE, E., and STORM, S. R. (1965), 'Cat heart muscle *in vitro*. VIII. Active transport of sodium in papillary muscles', *J. gen. Physiol.*, **48**, 957–972.

REFERENCES

PINSKER, H., and KANDEL, E. R. (1969), 'Synaptic activation of an electrogenic sodium pump', *Science, N.Y.*, **163**, 931–935.

POST, R. L., ALBRIGHT, C. D., and DAYANI, K. (1967), 'Resolution of pump and leak components of sodium and potassium ion transport in human erythrocytes', *J. gen. Physiol.*, **50**, 1201–1220.

POST, R. L., and JOLLY, P. C. (1957), 'The linkage of sodium, potassium, and ammonium active transport across the human erythrocyte membrane', *Biochim. biophys. Acta*, **25**, 118–128.

RANG, H. P., and RITCHIE, J. M. (1968), 'On the electrogenic sodium pump in mammalian non-myelinated nerve fibres and its activation by various external cations', *J. Physiol., Lond.*, **196**, 183–221.

RITCHIE, J. M., and STRAUB, R. W. (1956), 'The after-effects of repetitive stimulation on mammalian non-medullated fibres', *J. Physiol., Lond.*, **134**, 698–711.

RITCHIE, J. M., and STRAUB, R. W. (1957), 'The hyperpolarization which follows activity in mammalian non-medullated fibres', *J. Physiol., Lond.*, **136**, 80–97.

SCHATZMAN, H. J. (1953), 'Herzglykoside als Hemmstoffe fur den aktiven Kationaustausche an Rattenblutzellen', *Experientia*, **10**, 189–190.

SHAW, T. I., and NEWBY, B. J. (1970), personal communication.

SKOU, J. C. (1957), 'The influence of some cations on an adenosine triphosphatase from peripheral nerve', *Biochim. biophys. Acta*, **23**, 394–401.

SKOU, J. C. (1960), 'Further investigations on a Mg^{++} + Na^+ activated adenosinetriphosphatase, possibly related to the active, linked transport of Na^+ and K^+ across the nerve membrane', *Biochim. biophys. Acta*, **42**, 6–23.

SKOU, J. C. (1962), 'Preparation from mammalian brain and kidney of the enzyme system involved in active transport of Na and K', *Biochim. biophys. Acta*, **58**, 314–325.

SLAYMAN, C. L. (1965), 'Electrical properties of *Neurospora crassa*. Respiration and intracellular potential', *J. gen. Physiol.*, **49**, 93–116.

SLAYMAN, C. L. (1970), 'Movements of ions and electrogenesis in microorganisms', *Am. Zool.*, **10**, 377–392.

STEINBACH, H. B. (1940), 'Sodium and potassium in frog muscle', *J. biol. Chem.*, **133**, 695–701.

STRAUB, R. W. (1961), 'On the mechanism of post-tetanic hyperpolarization in myelinated nerve fibres from the frog', *J. Physiol., Lond.*, **159**, 19–20P.

STRUMWASSER, F. (1965), 'The demonstration and manipulation of a circadian rhythm in a single neurone', in *Circadian Clocks* (ed. ASCHOFF, J.), pp. 442–462. Amsterdam: North Holland Publishing Co.

TAMAI, T., and KAGIYAMA, S. (1968), 'Studies of cat heart muscle during recovery after prolonged hypothermia. Hyperpolarisation of cell membranes and its dependence on the sodium pump with electrogenic characteristics', *Circulation Res.*, **22**, 423–433.

TAYLOR, C. B. (1962), 'Cation stimulation of an ATPase system from the intestinal mucosa of the guinea pig', *Biochim. biophys. Acta*, **60**, 437–440.

TAYLOR, C. B., PATON, D. M., and DANIEL, E. E. (1969), 'Evidence for an electrogenic sodium pump in smooth muscle', *Life Science*, **8**, 769–773.

TEORELL, T. (1933), 'Untersuchungen über die Magensaftsekretion', *Skand. Arch. Physiol.*, **66**, 225–317.

THOMAS, R. C. (1968), 'Measurement of current produced by the sodium pump in snail neurone', *J. Physiol., Lond.*, **195**, 23–24P.

THOMAS, R. C. (1969), 'Membrane current and intracellular sodium changes in a snail neurone during extrusion of injected sodium', *J. Physiol., Lond.*, **201**, 495–514.

THURM, U. (1971), 'On the functional organization of sensory epithelia', *Verh. dt. zool. Ges.*, **64**, 79–88.

WHITTAM, R. (1964), *Transport and Diffusion in Red Blood Cells*. London: Arnold.

WHITTAM, R., and WHEELER, K. P. (1961), 'The sensitivity of a kidney ATPase to ouabain and to sodium and potassium', *Biochim. biophys. Acta*, **51**, 622–624.

WHITTAM, R., and WHEELER, K. P. (1970), 'Transport across cell membranes', *A. Rev. Physiol.*, **32**, 21–60.

WOOD, J. L., FARRAND, P. S., and HARVEY, W. R. (1969), 'Active transport of potassium by the *Cecropia* midgut. Microelectrode potential profile', *J. exp. Biol.*, **50**, 169–178.

AUTHOR INDEX

Adrian, R. H., 45–46, 111–114
Albright, C. D., 156, 159
Altamirano, M., 54
Alving, B. O., 44–45, 89–94, 103
Arvanitaki, A., 51–52
Ascher, P., 47, 50–51, 138, 141–144
Awad, M. Z., 111, 118, 158
Axelsson, J., 145

Baker, P. F., 20, 25
Baylor, D. A., 54, 148–152, 157
Bernstein, J., 26–27
Bohm, W. H., 65
Bonting, S. L., 159
Bos, C. J., 159
Bourke, R. S., 12
Boyle, P. F., 19
Brown, H. M., 145
Brown, L. C., 47–49, 138–142, 149
Bueding, E., 145
Bulbring, E., 145
Burnstock, G., 124

Caldwell, P. C., 20, 21, 24
Carey, M. J., 108
Carpenter, D. O., 44, 45, 89–94, 103
Casby, J. Y., 12
Chalazonitis, N., 51, 52
Clemons, R. D., 147
Cobb, D. M., 19
Connelly, C. M., 39, 40, 45, 62–64
Conway, E. J., 19, 104, 108, 157
Coombs, J. S., 45, 73, 85
Cordes, E. H., 24, 159–161
Cross, S. B., 111, 114–117, 119, 156

Daniel, E. E., 111
Dayani, K., 156, 159
Dean, R. B., 20
Desmedt, J. E., 115
Donnan, F. G., 14, 15, 17, 19, 21

Eccles, J. C., 45, 73, 85, 131
Eccles, R. M., 85
Eisenman, G., 12
Emmelot, P., 159

Farrand, P. S., 147
Fatt, P., 43, 73, 85
Fenn, W. O., 9, 19
Fingerman, M., 54, 152
Flear, C. T. G., 11
Florence, I., 11
Freygang, W. H., 113
Frumento, A. S., 111, 124–127

Garrahan, P. J., 156
Geduldig, D., 44, 111, 156
Gerard, R. W., 27
Gibbs, W. J., 14, 15, 17, 19, 21
Glynn, I. M., 22, 24, 25, 156
Godfraind, J. M., 144
Gola, M., 51, 52
Gorman, A. L. F., 81, 82, 83, 144
Graham, J., 27
Greengard, P., 40, 65, 70, 71
Grinnel, A. D., 70
Gruener, R., 70
Grundfest, H., 59, 99
Gurber, G., 6

Hagiwara, S., 70, 145
Haldane, J. S., 7–8
Hamburger, H. J., 6
Harris, E. J., 22
Harris, J. E., 22, 111, 114, 158
Harvey, W. R., 147
Hashimoto, Y., 111, 127–129
Haskell, J. A., 147
Hayashi, H., 70
Heppel, L. A., 19, 20
Herbert, N. C., 13
den Hertog, A., 70, 71
Hess, H. H., 159

AUTHOR INDEX

Hinke, J. A. M., 7, 8, 12
Hober, R., 19
Hodgkin, A. L., 20, 21, 27, 28, 29–32, 37, 38, 45, 57, 58, 112, 117, 155
Hoffman, J. F., 22
Hokin, L., 23
Holmes, O., 70
Horowicz, P., 28, 112, 117
Huxley, A. F., 27

Ito, M., 45, 73, 85

Jain, M. K., 24, 159–161
Jolly, P. C., 22, 156

Kagiyama, S., 45, 46, 121–124
Kandel, E. R., 50, 142–144, 145, 157
Kao, C. Y., 59
Kapitsa, P. C., 9
Kehoe, J. S., 50, 51, 141, 144
Kerkut, G. A., 40, 41–43, 44, 45, 47–49, 51–54, 74–79, 83–85, 138–142, 149
Kernan, R. P., 45, 108–111, 155, 158
Keynes, R. D., 11, 20, 21, 24, 27–32, 37, 38, 45, 57, 58, 61, 111, 114–117, 119, 155, 156
Kholodova, Yu. P., 13
Kobayashi, H., 137–138, 145
Koechlin, B. A., 11
Koike, H., 145
Koketsu, K., 46, 131–136, 137, 158
Kolmodin, G. M., 85
Kostyuk, P. G., 13
Krnjevic, K., 54, 144
Kuno, M., 74

Lavallee, M., 13
Lev, A. A., 12, 115
Lewis, P. R., 11, 20
Libet, B., 137–138, 145
Lloyd, P. C., 85
Løyning, Y., 85
Lundholm, L., 145

McConaghey, P. D., 22
Maizels, M., 22
Marmor, G., 81–83
Meech, R. M., 145

Michaelis, L., 26
Mirolli, M., 144
Miyahara, J. T., 74
Mohme-Lundholm, E., 145
Moreton, R. B., 79, 80
Mullins, L. J., 111, 118, 158
Murray, R. W., 44

Nakajima, S., 44–45, 95–99, 156–157
Nedergaard, S., 147
Netter, H., 15
Newby, B. J., 45, 155
Nicholls, J. G., 54, 148–152, 157
Nishi, S., 45, 46, 99–101, 131–137, 158

Obara, S., 99
Ochs, S., 111, 114, 158
Onodera, K., 54, 147
Oshima, T., 45, 73, 85
Overton, E., 6

Page, E., 111, 119–121, 124
Paton, D. M., 111
Patterson, J. H., 21
Pinsker, H., 50, 142–145, 157
Pope, A., 159
Post, R. L., 22, 156, 159
Priestley, J. G., 7
Pumain, R., 144

Rang, H. P., 40, 45, 65–69, 86, 110, 156–158
Ridge, R. M. A. P., 43, 44
Ritchie, J. M., 39, 40, 45, 59, 60–62, 70–71, 110, 156–158
Rudin, D. O., 12
Rybova, R., 111, 114–117, 119, 156

Sakata, H., 70
Salmoiraghi, G. C., 144
Schanne, O., 13
Schatzmann, H. J., 24
Shaw, T. I., 20–21, 45, 155
Skogland, C. R., 85
Skou, J. C., 23, 158, 159
Slayman, C. L., 45, 46, 105–107, 111–114
Soeda, H., 45, 99–101
Sorokina, Z. A., 13
Steinbach, H. B., 19

AUTHOR INDEX

Storm, S. R., 111, 119–121, 124
Straub, R. W., 39, 40, 45, 59–62, 65, 157
Strickholm, A., 24, 159–161
Strumwasser, F., 102
Swann, R. C., 61

Takahashi, K., 44–45, 95–99, 156–7
Tamai, T., 45–46, 121–124
Tangney, A., 110
Taylor, C. B., 111, 129, 159
Teorell, T., 16
Thomas, R. C., 41–43, 45, 47, 74–79, 83, 85–89, 156
Thurm, U., 146

Toteson, D. C., 22
Tower, D. B., 12

Walker, R. J., 47–49, 138–142, 149
Weakly, J. N., 74
Wheeler, K. P., 159
Whittam, R., 22, 159
Wood, J. L., 147

York, B., 51–54, 83–88

Zadunaisky, J. A., 108, 157

SUBJECT INDEX

A FIBRES in frog nerve, 62
Acetylcholine, 6, 47, 49, 50, 75, 76, 137, 138–145
— effect on snail neuron, 47–49, 130–142
Activity, 12
— potassium, in snail neuron, 13
— sodium, in snail neuron, 13
— synaptic, 131–138, 142–145, 148–152
— versus concentration, 12
Adaptation, 6, 54
Adenosine triphosphatase (ATPase), 20, 23, 24, 33, 158–161
— — of electrogenic sodium pump, 20–24, 158–161
— triphosphate (ATP), 4, 20, 29, 158, 161
Adrenaline, 131, 145
Alloemicymarin, 24, 25
Amytal, 114
α-Angelica lactone, 25
Anisdaris nobilis, 144–145
Anoxia, 6, 51–54, 65, 83–85, 137, 147
— electrogenic sodium pump in, 51–54, 83–85
Antimycin A, 40
Aplysia, 44, 45, 47, 50, 51, 52, 89–94, 102, 103, 138, 141–144
— nerve, electrogenic sodium pump in, 41–45, 89–94, 141–144
Archidovis nobilis, 82–83
Artificial membranes, 159–161
ATPase (see Adenosine Triphosphatase)
Axon, squid, 11, 12, 20, 21, 24, 27, 45, 155, 157–159
Azide, 20, 29, 31, 40, 46, 103–105, 137

B FIBRES in frog nerve, 59
Barium, 99–101
Barnacle photoreceptor, electrogenic sodium pump in, 145
Black lipid membranes (BLM), 159–161

Blood-corpuscle, red, 10, 21, 24, 25, 156
Br neuron of *Aplysia*, 138

C FIBRES, 59, 163
Caesium, 69
Calcium, 24
Carbon monoxide, 103
Carbonic anhydrase, 8
Cardiac glycosides, 24, 25, 26
Cat cortical neurons, 54
— heart, electrogenic sodium pump in, 119–122
— — muscle, 45, 46, 119
Cecropia midgut, 147
Cells, glial, from leech (*Hirudo*), 54, 148–152, 157
— pigment, 152–153
Cervical sympathetic nerve-trunk, 39, 59–62
Chloride, 17, 18, 26, 40, 47, 49, 55, 58, 65, 66, 75, 76, 82, 100, 102, 103, 109, 110, 113, 115, 118, 130, 140, 141, 143
— ions, diffusion of, 17, 18
Cholesterol, 159
Choline, 69
Chromatophores, 6, 54, 152–153
CILDA, 47, 48
Clamp, current, 43, 85–87
Classic view of membrane potential, 33, 34, 138, 140
Cocaine, 93, 94, 113, 114
Collodion, 26
Concentration versus activity, 12
Conclusions, 163
Constant-field equation, 80, 83
Cortical neurons, 12, 54
Crab muscle, 43
— nerve, 11
Crayfish stretch receptor, 44, 45, 54, 95–99, 148, 156
Curare, 50, 137, 141, 143
Current clamp, 43, 85–87
— short-circuit, 5, 65–69, 110

SUBJECT INDEX

Cyanide, 20, 29, 30, 35, 40
Cyclical theory, 1
Cytochrome(s), 104
— c reductase, 40

DEMARCATION potential, 34
Denervation, 110
Deoxy-D-glucose, 70–71
Differential membrane solubility, 16
Diffusion of Cl⁻ ions, 17, 18
— H⁺ ions, 15, 17
Digitonin, 24
Digoxin, 24, 25
DILDA, 138
2,4-Dinitrophenol (DNP), 20, 29, 30, 31, 32, 35, 39, 40, 46, 58, 60, 61, 95, 96, 97, 103–105, 127, 128, 137, 144, 147
Dog nerve, 11
Dopamine, 47, 138, 139, 142
— effect on snail neuron, 48, 138–142

E_{Cl}, 75, 76, 133, 138
E_{K}, 5, 33, 34, 35, 36, 39, 40, 62, 73, 79, 94, 97, 98
E_{Na}, 107, 108, 110, 114, 115, 116, 117, 119, 122, 127, 128, 129, 133, 134, 145
Electrode(s), glass, 10–13
— — potassium-sensitive, 12, 13, 15
— sodium-sensitive, 12–13, 43, 85–89
Electrogenic sodium pump in anoxia, 51–54, 83–85
— — — Aplysia nerve, 44–45, 89–94, 141–144
— — — ATPase of, 20–24, 158–161
— — — in barnacle photoreceptor, 145
— — — black lipid membrane and, 159–161
— — — in cat heart, 119–122
— — — definition of, 5, 35, 36, 37, 55
— — — in frog muscle, 62, 108–109
— — — — nerve, 62–65
— — — — spinal cord ganglion, 99–101
— — — — sympathetic ganglion, 46
— — — leech (Hirudo) glial cells, 54, 148–153, 157

Electrogenic sodium pump and membrane potential 37
— — — in nerve axon, 28, 57–71
— — — nerve-cells, 73–104
— — — and post-tetanic hyperpolarization (PTH), 38, 59–71, 95–99
— — — in snail (Helix) neurons, 41–44, 74–89
— — — sodium–potassium linkage in, 4, 20, 22, 29, 32, 83, 151–152, 155–157
— — — temperature-sensitivity of, 43, 89–94, 119–128
— — — versus electroneutral pump, 32–33, 61
Electroneutral pump, 32–33, 61
Emicymarin, 24, 25
Energy barrier, 108, 157–158
Epithelial potential, 146
Epithelium, sensory, 146
EPSP (see Excitatory Post-synaptic Potential)
Equilibrium, Gibbs-Donnan, 14, 15, 17, 18, 21
Ergometrine, 142
n-Ethyl maleimide (NEM), 24
Excitatory post-synaptic potential (EPSP), 46, 131, 135, 137, 145, 149

FIBRE(s), A, in frog nerve, 62
— B, in frog nerve, 59
— C, 59, 163
Flame photometer, 9, 108
Fluoride, 22
Frog muscle, 43, 45, 107, 108, 109–119, 124–129
— — electrogenic sodium pump in, 62, 108–109
— nerve, electrogenic sodium pump in, 62–65
— peripheral nerve, 31
— spinal cord ganglion, electrogenic sodium pump in, 99–101
— sympathetic ganglion, 46

GAMMA-AMINO butyric acid (GABA), 144
Ganglia, frog spinal cord, 99–101
— — sympathetic, 46
— sympathetic, 46, 131–138, 158
Generator potential 6

SUBJECT INDEX

Ghosts, 22
Gibbs-Donnan equilibrium, 14, 15, 17, 18, 21
Glass electrode(s), 10–13
— — potassium-sensitive, 12, 13, 15
Glial cells of leech, 6, 54, 148–152, 157
Glucose, 22, 70, 71
Glycosides, cardiac, 24, 25, 26
Golf-ball typewriter, 3

HEART, cat, electrogenic sodium pump in, 119–122
— muscle from cat, 45, 46, 119
Helix neurons, electrogenic sodium pump in, 41–44, 74–89
Henry's law, 8
Hexahydroscillaren, 24, 25
Hirudo glial cells, 54, 148–152, 157
Hydrogen ions, diffusion of, 15, 17
Hyperpolarization, post-tetanic (PTH), 30, 39, 40, 44, 45, 59–71, 74, 95–99, 156

IBM, 3
Inhibiting post-synaptic potential (IPSP), 149, 150
Inhibitors, metabolic (*see also* Azide, Cyanide, 2,4-Dinitrophenol), 70
Injection of sodium, 33, 34, 36, 37
Insect muscle, 43
Insulin, 109
Inulin space, 11
Iodoacetate, 40, 47, 127, 128
Isethionate, 40, 66, 67

JUNCTION potentials, 35, 40

LACTATE, 109
Leech (*Hirudo*) glial cells, 54, 148–152, 157
Leptodactylus ocelatus, 124
Lithium, 61, 62, 69, 73, 76, 91, 92, 95, 96, 99, 111, 118, 119, 123
— effect on snail neuron, 76
Lobster nerve, 11

MAGNESIUM, 4, 23, 24
Magnetic tape recording, 3
Malonate, 71

Membrane(s), artificial, 159–161
— black lipid (BLM), 159–161
— potential, classic view of, 33, 34, 138, 140
— — and electrogenic sodium pump, 37
— resistance, 5, 113
— solubility, differential, 16
Mepyramine, 114
Metabolic inhibitors (*see also* Azide, Cyanide, 2,4-Dinitrophenol), 70
Midgut of *Cecropia*, 147
Mitochondria, 26
Motoneuron, 73–74, 122
Musca, 146
Muscle, 10, 27, 28, 45, 107, 122
— cat heart, 45, 46, 119
— crab, 43
— frog, 43, 45, 107, 108, 109–119, 124–129
— — electrogenic sodium pump in, 19–20, 62, 108–109
— insect, 43
— rat, 19
— smooth, 129
Myometrium, 129

NERNST equation, 27, 28, 34, 37, 43, 46
Nerve(s), *Aplysia*, 41–45, 89–94, 141–144
— axon, electrogenic sodium pump in, 28, 57–71
— crab, 11
— dog, 11
— frog, electrogenic sodium pump in, 62–65
— — peripheral, 31
— lobster, 11
— rat, 11
— sciatic, 110
— sodium pump in, 20
Nerve-cell, electrogenic sodium pump in, 73–104
Nerve-trunk, sympathetic cervical, 39, 59–62
Neuroglia, 144, 148–152
Neuron(s), Br, of *Aplysia*, 138
— cat cortical, 54
— cortical, 12, 54
— snail (*Helix*), 41–44, 74–89
Neurospora, 45, 103–105
Nitella, 104
Nitrogen, 51–54, 83–85, 103
Noradrenaline, 123, 124, 137, 142

SUBJECT INDEX

OCTYLGUANIDINE, 24
Oligomycin, 24
Ouabain, 5, 24, 26, 39, 41, 42, 46, 49, 53, 61, 66, 67, 68, 77–78, 81, 85, 90, 91, 99, 111, 116, 123, 127, 128, 129, 137, 138, 144, 145
— effect on snail neuron, 77
Oxygen, 6, 51
— tension, effect on snail neuron, 8, 51, 83–85

P POTENTIAL, 46, 131–138, 158
Parabolic burster, 102–103
Parachloromercuribenzoate, 24, 77, 78
Peripheral nerve, frog, 31
pH, 16, 70
Phosphorylase, 145
Photometer, flame, 9, 108
Photoreceptor of barnacle, 145–146
Physics and biology, 8, 89
Piezo-electric effect, 148
Pigment cells, 152–153
Post-synaptic potential, excitatory, 46, 131, 135, 137, 145, 149
— — inhibitory (IPSP), 149, 150
Post-tetanic hyperpolarization (PTH), 30, 39, 40, 44, 45, 59–71, 74, 95–99, 156
— — and electrogenic sodium pump, 38, 59–71, 95–99
Potassium (see E_K)
— acetate injection, effect on snail neuron, 41–42, 74–75
— activity in snail neuron, 13
Potassium-sensitive glass electrodes, 12, 13, 115
Potential(s), demarcation, 34
— epithelial, 146
— excitatory post-synaptic (EPSP), 46, 131, 135, 137, 145, 149
— generator, 6
— inhibitory post-synaptic (IPSP), 149, 150
— junction, 35, 40
— membrane, classic view of, 33, 34, 37, 138, 140
— P, 46, 131–138, 158
— receptor, 44, 45, 54, 95–99, 146–148, 156
— sensory, 148
Procaine, 114
Progress, scientific, 1

Pump current in snail neuron, 43, 85–89
Pyruvate, 22, 71

RAT muscle, 19
— nerve, 11
Receptor potential, 44, 45, 54, 95–99, 146–148, 156
— stretch of crayfish, 44, 45, 54, 55, 95–99, 147, 148, 156
Red blood-cells, sodium pump in, 21–22
— blood-corpuscle, 10, 21, 24, 25, 156
Resistance of membrane, 5, 113
Ringer, sulphate, 109, 118
Rubidium, 46, 69, 73, 112, 113

SCIATIC nerve, 110
Scientific progress, 1
Scillaren A, 24, 25
Sensory epithelium, 146
— potential, 148
Short-circuit current, 5, 65–69, 110
Smooth muscle, 129
Snail neurons, constant field equation and, 79–81
— — effect of acetylcholine, 47–49, 138–142
— — — dopamine, 48, 138–142
— — — lithium, 76
— — — ouabain and K-free Ringer, 77
— — — oxygen tension, 8, 51, 83–85
— — — sodium and potassium acetate injection, 41–42, 74–75
— — — temperature, 43
— — electrogenic sodium pump in, 41–44, 74–89
— — potassium activity in, 13
— — pump current in, 85–89, 43
— — sodium activity in, 13
— — type I to ACh, 138–141
— — — II to ACh, 138–141
Sodium acetate injection, effect on snail neuron, 41–42, 74–75
— activity in snail neuron, 13
— methyl sulphate, 110
Sodium–potassium ATPase, 20, 23, 24, 33, 158–161

SUBJECT INDEX

Sodium–potassium linkage in sodium pump, 4, 20, 22, 29, 32, 83, 151–152, 155–157
— pump, 4–5, 19–22
— — in muscle, 19–20
— — nerve, 20
— — red blood-cells, 21–22
Sodium-sensitive electrodes, 12–13, 43, 85–89
Solubility, differential membrane, 16
Squid axon, 11, 12, 20, 21, 24, 27, 45, 155, 157–159
Straight-line theory of progress, 1
Stretch receptor of crayfish, 44–45, 55, 95–99, 147–148, 156
Strophanthin, 24, 148, 149
Sucrose gap, 40, 46, 59, 65, 140
Sulphate Ringer, 109, 118
Swim-bladder, 7
Sympathetic ganglion, 46, 131–138, 158

Sympathetic nerve-trunk, cervical, 39, 59–62
Synaptic activity, 131–138, 142–145, 148–152
Synaptosomes, 24, 160

TEMPERATURE, 43, 44, 46, 74, 81, 82, 89–94, 101, 112, 116, 119–125, 129, 136, 142, 149, 153
— effect on snail neuron, 43
Temperature-sensitivity of sodium pump, 43, 89–94, 119–128
Tetrodotoxin, 47, 97, 103, 141–142
Thallium, 69
Typewriters, 3

VITAL force, 23
Vitalist, 8